Forest
and
Civilisations

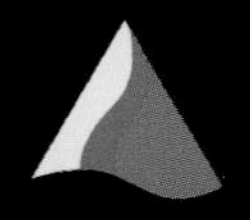

YANGTZE RIVER CIVILISATION PROGRAMME
(YRCP)

Forest and Civilisations

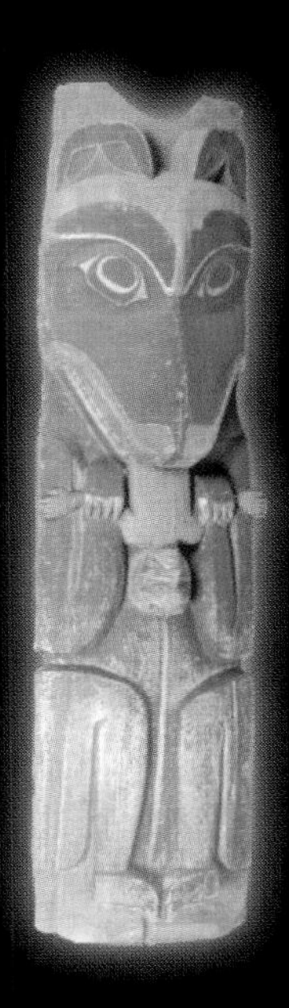

EDITED BY
Yoshinori Yasuda
International Research Center
for Japanese Studies

Lustre Press
Roli Books

The Grant-in-aid Programme for COE Research Foundation of the
Ministry of Education, Science, Sports and Culture in Japan
A Study of the Yangtze River Civilisation

ISBN : 81-7436-145-6

3-2 Oeyama-cho, Goryo, Nishikyo-ku, Kyoto 610-1192, Japan
Tel: 81- (75) 335-2150, Fax: 81- (75) 335-2090
E-mail: yasuda@nichibun.ac.jp, Website: http://www.nichibun.ac.jp

First published in 2001 by
Roli Books Pvt. Ltd.
Lustre Press Pvt. Ltd.
M 75 Greater Kailash II Market, New Delhi 110 048, India
Tel: (011) 6442271, 6462782, Fax: (011) 6467185, 6213978
E-mail: roli@vsnl.com, Website: Rolibooks.com

EDITOR:
Yoshinori Yasuda

ASSOCIATE EDITORS:
Haruno Ogasawara, Vasant Shinde

PHOTOGRAPHS:
Yoshinori Yasuda
Takao Inoue
Hiroko Yoshino
Hillel Burger, Peabody Museum of Archaeology and Ethnology,
Harvard University, Cambridge
American Museum of Natural History, New York
Thomas Burke Memorial Washington State Museum, Seattle

Produced at Roli CAD Centre

Printed and bound in Singapore

To Takeshi Umehara, former director general of International Research Center for Japanese Studies—one of the first who pointed out the importance of the forest culture philosophy for the past and the future world.

Contents

Tales of the Forest

YOSHINORI YASUDA

Under the Gaze of the Forest Gods

Once, our ancestors lived their lives surrounded by forests, which provided them abundant gifts of nature. These forests were home to the gods. Aware of the eyes of these gods continually watching them, humans built statues of these awesome gods with enormous eyes.

We see them in the eyes of the Medusa that appears in Greek mythology, the eyes of the Moai monoliths on Easter Island, and the eyes of clay figures (*dogu*) that date back to the Jomon Period in Japan. All these eyes were once those of the forest gods keeping a careful watch as our ancestors went about their daily affairs.

After a certain point in time, however, humans stopped making statues of gods and even started destroying them. They stopped making Jomon clay figurines, toppled the Medusa from the girders of the shrine, and knocked down the Moai of Easter Island.

Instead of worshipping the forest gods, humans now made the living creatures that dwelt within the forests into demons and witches. Then, humans, in pursuing the demons and witches, cut into the forests, only to cause intense devastation and disappearance of the forests.

Forgetting that they had been blessed by the forests, humans came to regard the forest with insolence. They thought that they had been successful in creating a kingdom of humans on earth. Their illusion continued until the end of the twentieth century. The realisation is dawning, if only painfully, slowly that the only way left for humans to continue to live their lives is to recapture a sense of reverence toward nature and the earth by recognising that they are watched over and protected by the forest.

Is 'History' Narrated Only From the Perspective of the Victors?

We have been taught 'history' from the perspective of the victors of modern European civilisation that propped up the spiritual supports for Christianity. It was this civilisation that turned Medusa, who was once a god, into a monster. The modern Europeans also buried the civilisations of the Indio, the original inhabitants of central and south America, and the Indians (the native Americans) of north America into the dark corners of history by labelling them barbarians. Is it not true that the world history we have learned is a history that supports the global dominance of European civilisation?

In today's world it goes without saying that nature is the weakest organism, along with the forests and the living creatures that dwell within them. It is necessary today that a world history be written from the perspective of the weak; of those who have been controlled and exploited. What must be exposed are the truths that have been buried in our conventional understanding of a world history.

This book is an attempt to reinterpret world history and present the stories of the relationship between the forest and human beings from the viewpoint of environmental archaeology as a field of historical earth science. In the field of the natural sciences, story telling has been taboo; to say that one's research paper is merely a story implies that the quality of research is dreadful. When discussing the relationship between forests and humans, however, it is impossible to do so without telling these stories. Regardless of the advances that have been made in botany and zoology, natural scientists have been unable to prevent the destruction of forests and the extermination of many animal species.

The reason for this is that the natural sciences have until now lacked a sense of narrative. In place of the legacy of modern European science, new stories must now be told; new stories which take the position of the weakest subjects on earth. In other words, the stories must be told from the perspective of both the forests that continue to be destroyed and eradicated, and the minority indigenous peoples and animals that dwell within them. Such stories will help us to re-examine how science is undertaken today and to produce a movement aimed at creating a new science of the relationship between nature and human beings.

Moais are standing at Tahai watching over inland. Photo by Yasuda

Part I

CHANGES IN THE FOREST AND CIVILISATIONS

CHAPTER 1

Comparative Study of the Myths and History of a Cedar Forest Each in East and West Asia

YOSHINORI YASUDA

The Spirit of the Forest in the Epic of Gilgamesh

■ GILGAMESH AND HUMBABA

The epic of Gilgamesh is the oldest written tale. Gilgamesh (Fig. 1), the hero of the epic, is taken to have been a real king of the City of Uruk (Fig. 2) around 2600 BC. The king was made a god following his death, and his many heroic exploits are recounted in Sumerian. The epic is believed to have been completed at the very latest by the end of the New Sumerian Period, around 2600 BC.

The epic was thereafter recounted in Mesopotamia and on the Anatolian plains, and underwent slight changes each time. It was also passed on during the Babylonian era, when Akkadian was the common language of Mesopotamia, and has also been found written in Hittite and Hurrian, at Bokazkoy, the capital of the Hittite kingdom on the Anatolian plains.[1] The hero of the epic, Gilgamesh, is called Bilgamesh in Sumerian, and Humbaba is called Huwawa. The names, Gilgamesh and Humbaba, are derived from Akkadian.

The ancient Babylonian version of Gilgamesh, written around 2000 BC, says the following: Gilgamesh cried, 'I shall kill Humbaba, whom I ought to fear, I shall vanquish evil, and will cut down the Lebanese cedar.'[2] His friend Enkidu (Fig. 3) and the elders, seeing this, tried to stop him from entering the forest, saying, 'The forest

Fig. 1: King Gilgamesh, mural carved in the Cholsavad Palace, Assyrian Empire in Iraq (Courtesy Louvre, Paris. Photo by Yasuda)

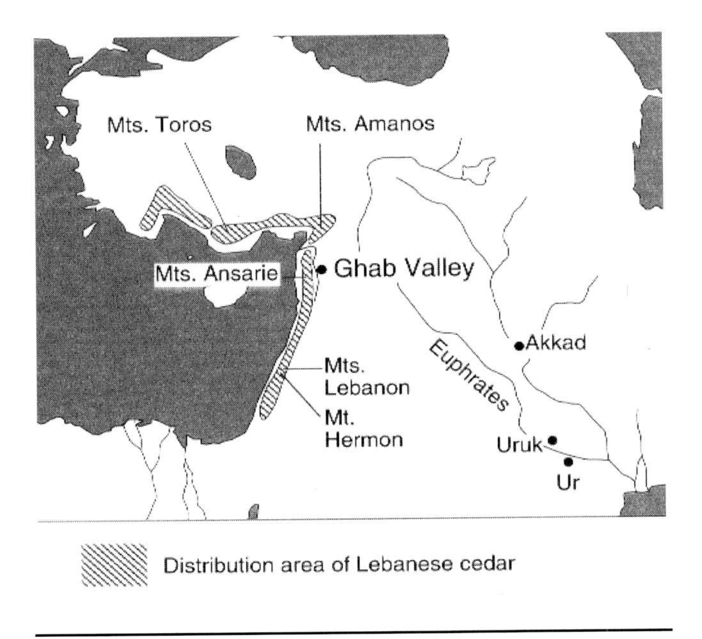

Fig. 2: The lands of the epic of Gilgamesh and the distribution of Lebanese cedars

spreads for 10,000 leagues, and no lone soul can approach. The voice of Humbaba can raise great floods, and fire and death mix in his breath.'[3]

However, as if to quell this, Gilgamesh retorted, 'I want to see Humbaba, god of the forest of whom people speak. I shall conquer Humbaba in the forest of Lebanese cedars. Then I shall cut down the cedars, to create a name for myself that shall last for all eternity'.[4] So saying, Gilgamesh and Enkidu reached the forest of Lebanese cedars (Fig. 4): 'Taking his axe in his hand, and unsheathing his giant sword from upon his hip, Gilgamesh struck Humbaba's head. Enkidu struck at his heart. With the third blow, Humbaba was felled.'[5] 'Gilgamesh grasped Humbaba's head and knocked it in with his mallet.'[6] With the death of Humbaba, peace returned to the forest. Saria (Hermon) and the mountains of Lebanon shook with the roars of Gilgamesh and Enkidu.'[7] The bandits of the forest had been destroyed, and Gilgamesh cut down the mountains of Lebanese cedar five times, while Enkidu dug up their stumps. Then, they carried the cedars to the temple of Enlil across the Euphrates on a raft.

The tale suddenly develops into a saga of Gilgamesh's quest for eternal life after he loses his friend Enkidu. By the sea, Gilgamesh asks to be shown the path to a meeting with Utnapishtim. The water maiden introduces Gilgamesh to Sursunabu, servant of Utnapishtim, but the sequence of the tale is missing after they crossed the Sea of Death by boat.

It is fair to view the ancient Babylonian version of the epic as being the closest to that passed down from Sumeria. In the ancient past of 2600 BC, why did the people of Mesopotamia choose to recount an epic whose principal theme is the killing of the god of the forest; in other words, the destruction of the forests of Lebanese cedar? If this were a heroic legend, it would have been better to recount military glory arising from wars between magnificent city states, but the author of this epic poem chose the battle with Humbaba, god of the forest, as its central topic. Why did the

Fig. 3: Gilgamesh (centre) and Enkidu (Allepo Archaeological Museum. Photo by Yasuda)

Fig. 4: The forests of Lebanese cedar were Humbaba's home, Jadji in Lebanon (Photo by Yasuda)

author of Gilgamesh select a tale of the destruction of the forests as the epic's central motif in 2600 BC?

The standard version of the epic, written some 900 years after the ancient Babylonian one, is the most widely known today and is also very well preserved. This standard version of the epic of Gilgamesh was established around 1100 BC. Over the passage of these 900 years, considerable changes in expressions of the epic can be identified.

First, the splendour of the Lebanese cedar forests where Humbaba dwells (as Huwawa is called in the standard version) is emphasised. There are many gaps missing from the ancient Babylonian version, but the line 'They gazed motionless upon the forest' still remains. In the

Fig. 5: The forests of Lebanese cedar, enveloped in pure white snow in Busshari, Lebanon (Photo by Yasuda)

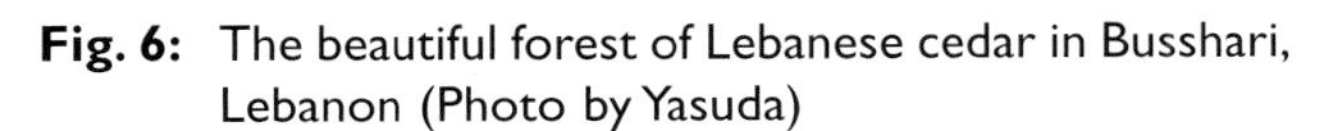

Fig. 6: The beautiful forest of Lebanese cedar in Busshari, Lebanon (Photo by Yasuda)

Fig. 7: Straight trunks of the Lebanese cedar stand like the pillars of Humbaba's temple (Photo by Yasuda)

standard version, however, we find that not only is the excellence of the literary expression praised, but also the beauty of the forest (Fig. 6 and 7): 'Gilgamesh and Enkidu froze and stared into the woods' great depth and height. When they espied Humbaba's path, they found the opening toward a straight passage. Then they were able to find and see the home of the gods, the paradise of Ishtar's other self, called Imini (most–attractive). All beauty true is ever there where gods do dwell, where there is cool shade and harmony and sweet-odoured food to match their mood.'[8] This is not only due to an improvement in literary expression over the passage of 900 years, but also because of changes in the way people viewed the Lebanese cedar forests.

The second major difference is that whereas the standard version emphasises the payback for the sins Gilgamesh committed, the Babylonian version does not. Gilgamesh, after having killed Humbaba, spurns the advances of Ishtar, the earth goddess. Angered, Ishtar asks the god Anu for the Bull of Heaven, in an attempt to destroy Gilgamesh and Enkidu. The Bull of Heaven is a giant creature that 'cracks the earth to swallow up nine dozen citizens of Uruk! An earthquake fixed a grave for nine dozen citizens of Uruk. Two hundred victims, maybe more than that, fell into Hell.'[9]

In ancient Mesopotamia, the bull (Fig. 8) was the largest beast. Such a fearsome life force was the symbol of an abundant harvest. The bull's testicles in particular were worshipped as a symbol of reproduction and fertility. Faith in the bull such as this has been widely identified from ancient Mesopotamia to the Anatolian plains, as well as in the ancient Mediterranean world. The Minotaur of the island of Crete is another manifestation of this faith in the bull. Enkidu also falls into a pit dug by the Bull of Heaven, but he jumps out and grabs the Bull by the horns. Gilgamesh stabs the Bull to death with his sword, and cuts out its heart.

However, because he has slain Humbaba and the Bull of Heaven, Enkidu is singled out by the gods for death. In the standard version of the epic of Gilgamesh, the second point that is emphasised is the payback for the sins Gilgamesh has committed: the killing of the Bull of Heaven and of Humbaba which is a symbol of the bounteous harvests of Mother Earth. Gilgamesh's suffering at his friend having been marked for death is profound, and he pleads with the gods for Enkidu's life. However, Enkidu's illness is severe, and his body grows weak. His death comes when 'Once again at break of day did Gilgamesh conclude the silent night.'[10] Gilgamesh sings a long eulogy at the death of his friend: 'We went up on mountains high to where we dared to capture god's own strength in one great beast and then to cut its throat, thus humbling Humbaba, green god of woodlands steep. Now there is a sleep-like spell on you, and you are dark as well as deaf. Enkidu can move no more. Enkidu can lift his head no more. Now there is a sound throughout the land that can mean only one thing. I hear the voice of grief and I know that you have been taken somewhere by death.'[11]

Faced with the death of his friend, Gilgamesh is as agitated as a lioness whose cub has been taken, and tears out his hair. He then embarks on a long voyage to cast away the fear of death and resolve the riddles of life and death. This is recounted as, 'When day did break again next morning, Gilgamesh stripped off the lion's cloak and rose to say this prayer.'[12]

The standard version of the epic of Gilgamesh, written around 1100 BC, still tugs at the hearts of modern people after 3,000 years. When the first rays of dawn flickered in Mesopotamia and the Mediterranean world, this was the most beautiful moment of the day, along with the setting of the sun. The author of 'Gilgamesh' depicts the pain Gilgamesh experiences when he leaves following the death of Enkidu at this beautiful moment as if it were yesterday.

However, although the main storyline of the ancient Babylonian version created around 2000 BC and the standard version are very similar, we can see subtle changes in the focus placed by the authors.

After the passage of 900 years, people felt a strong affinity to the beauty of the forests of Lebanese cedar. They had become far more strongly aware of the meaning of the retaliation of the gods for the killing of Humbaba, god of the forest, and the killing of the Bull of Heaven, which was the messenger of the gods of Mother Earth. In all likelihood, what the author of the epic wanted to point out when describing the death of Enkidu and the grieving of Gilgamesh is the pain of absolution suffered for killing the gods of nature.

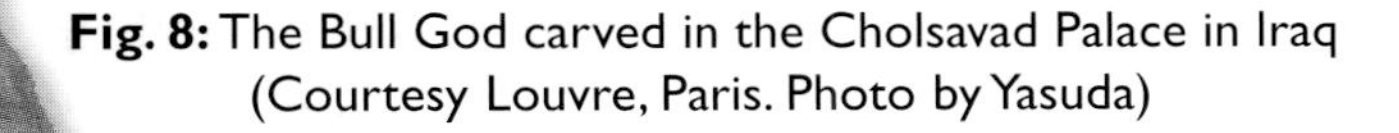

Fig. 8: The Bull God carved in the Cholsavad Palace in Iraq (Courtesy Louvre, Paris. Photo by Yasuda)

Fig. 9: The boring site at Ghab Valley against Mt Ansarie in north-west Syria (Photo by Yasuda)

What then must have led to such changes? The results of our pollen analysis from Ghab Valley in north-western Syria have yielded one potential clue to the solution of the mystery of this ancient myth.

■ RESULTS OF POLLEN ANALYSIS FROM GHAB VALLEY

Reaching 240 to 300 m above sea level, Ghab Valley is a rift valley located in the middle reaches of the Orontes River in north-western Syria (Fig. 2). The region used to be called Lake Nia, and was famous for elephant hunting around 1450 BC, during the reign of the Pharaoh Thutmes III of Egypt. The region has a Mediterranean climate, with rain in winter. The mean annual temperature of Idrib, located 30 km north-east of Ghab Valley, is 17.9°C, and the mean annual precipitation 429 mm. The region is linked by Mt Ansarie rising to the west of Ghab Valley to a height of approximately 1,300 m.

The western slope of Mt Ansarie descends relatively gently to the Mediterranean, but the eastern face falls sharply to Ghab Valley due to a fault line. Up to 800 m above sea level on these slopes formed of Tertiary Miocene limestone cover maquis which grow *Pinus brutia*, *Quercus calliprinos*, and *Crataegus monogyna*. Olives are also cultivated there. Above 800 m grow stunted broad-leaved deciduous trees, mostly deciduous oaks, such as *Q. aegilopus*, *Q. infectoria*, and *Q. cerris*. Above 1,000 m, the tall Lebanese cedar grows sparsely, occasionally dotting the landscape. In 1991, the Grant-in-Aid for Scientific Research on Priority Areas from the Japanese Ministry of Education, Science, Sports and Culture in Japan enabled us to collect continuous core for pollen analysis in the south-west of Ghab Valley (35° 39' N., 36° 15' E., at a height 240 m above sea level) (Fig. 9).

Fig. 10 shows a cross section of 6 m of sediment. The sediment from the bottom up is as follows: clay, organic clay, silty clay, and fine sand. In between are sandwiched several layers of terrestrial shells. Dr Hiroyuki Kitagawa of the International Research Centre for Japanese Studies (JAS) used a liquid scintillation counter on the seventh horizon of these shells to measure

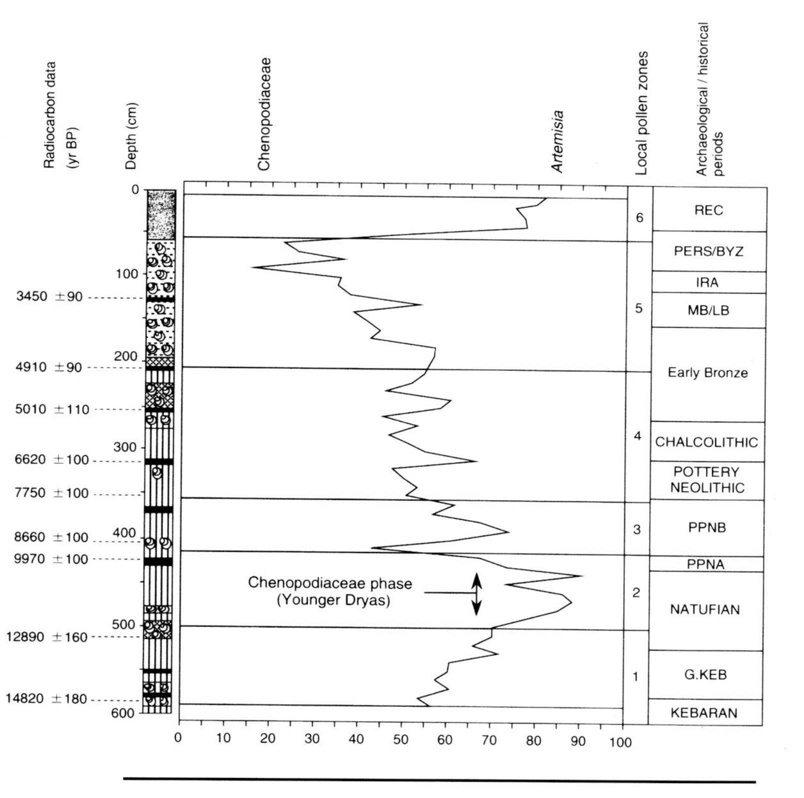

Fig. 10: Site stratigraphy and Artemisia/ Chenopodiaceae ratio with its chronological correlation to regional archaeological/ historical periods (Yasuda, et al., 2000). Archaeological/ historical periods were defined by Baruch and Bottema (1999)
G. KEB: Geometric Kebran; PPNA: Pre-Pottery Neolithic A; PPNB: Pre-Pottery Neolithic B; MB/LB: Middle Bronze Age/Late Bronze Age; IRA: Iron Age, PERS/BYZ: Persia-Byzantine; REC: Recent

their ^{14}C date. He also used an accelerator (AMS) at Nagoya University on the second horizon. The results of the ^{14}C dating clearly show that the 6 m of sediment cover a record of at least 15,000 years.

The method of pollen analysis is as follows: Hydrochloric acid processing → washing with water → KOH processing (warmed by steam bath for 10 min. in 70% potassium hydroxide) → washing with water → gravity separation (using 70% zinc chloride solution, gravity 2.0) → washing with wateracetic acid processing → acetolysis processing (warmed by steam bath for 3 minutes in a liquid mixture of 9 parts acetic anhydride to 1 part concentrated sulphuric acid) → acetic acid processing → washing with water → and then viewing under a microscope in 50% glycerine. Using the microscope, over 200 tree pollen grains (Fig. 11) were identified.

The results of the pollen analysis are shown in Fig. 12 and Fig. 13. The relative pollen diagrams in Figs. 12 and 13 were constructed in terms of percentages to the total pollen sum, excluding aquatic pollen. The pollen diagrams were classified by local pollen assemblage zones 1 to 6 from the lowest upwards.

Local pollen assemblage zone 1 is equivalent from 14,500 to 12,500 ^{14}C years BP. The results showed high concentrations of goosefoot (Chenopodiaceae) and mugwort (*Artemisia*), and showed that at the end of the last-glacial maximum, dried goosefoot and mugwort plains had spread across Ghab Valley. However, deciduous oak pollen had also increased towards the top end of assemblage zone 1, showing that forests of deciduous oaks had begun to gradually expand throughout the locality.

Local pollen assemblage zone 2 is equivalent from 12,500 to 9,000 ^{14}C years BP. During this period, the mugwort declined rapidly, and the deciduous oak pollen appeared in greatest numbers across all the horizons. Along with the end of the last-glacial period, the temperature and humidity also rose, the grass plains shrank in

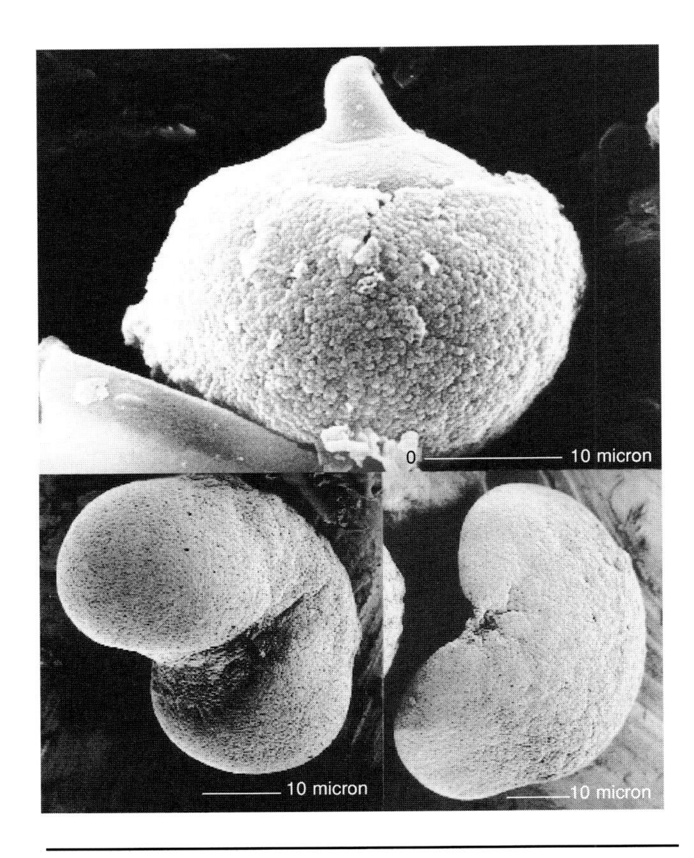

Fig. 11: Scanning Electron Microscopic photographs of fossil pollen
1. Japanese cedar (*Cryptomeria japonica*), 2 and 3. Lebanese cedar (*Cedrus libani*). Bar = 10 µm.

extent, and the forests of deciduous oak expanded locally.

Local pollen assemblage zone 3 is equivalent from 9,000 to 7,700 ^{14}C years BP. There are major changes in this horizon, which yielded 8,655±95 ^{14}C years BP. From approximately 30%, the deciduous oak pollen suddenly drop to less than 10%. Instead, there is a rapid increase in charcoal fragments, which speaks of human activity. There is also an increase in olive (*Olea*) pollens. After a slight delay, there is also a rapid increase in pine (*Pinus*) pollens. These clearly tell a tale of the destruction of the deciduous oak forests on the slopes of Mt Ansarie due to such Pre-Pottery Neolithic B (PPNB) people's activity, such as the cultivation of wheat and olives, and the growth of pine trees thereafter as secondary forest.

Local pollen assemblage zone 4 is equivalent from 7,700 to 4,900 ^{14}C years BP. In the horizon that yielded 7,750 ± 100 ^{14}C years BP, this time we see a rapid decrease in pine and Lebanese cedar (*Cedrus*). Alongside the pines, which increased as secondary forest, there are also indications that the Lebanese cedar, which grew in higher places above sea level, was destroyed 7,700 years ago. It is surprising that the massive clearance of Lebanese cedar forests occurred over 3,000 years before the appearance of King Gilgamesh.

Local pollen assemblage zone 5 is equivalent from 4,900 to 1,000 ^{14}C years BP. In the horizon that yielded 4,910±85 ^{14}C years BP, the deciduous oak also finally reduces to less than 5%. With the exception of maquis which grow *Quercus calliprinos* and olives, nearly all the forests have disappeared from the slopes of Mt Ansarie and from this we can thus deduce that the forests on Mt Ansarie were totally destroyed. Once this era began, bulrushes (*Typha*) increase their numbers rapidly. Instead of the comparatively dry goosefoot and mugwort plains to date, there is clear indication of the spread of wetland plains such as bulrushes and umbrella plant (Cyperaceae). The expansion of these bulrush wetlands reflects climatic changes around 5,000 ^{14}C years BP. in the Near East that have been pointed out in Yasuda (1990, 1991, 1995).

In local pollen assemblage zone 6, pines once again begin to expand, which tells us of a re-expansion of pine as a secondary forest on Mt Ansarie. This period is the time when vegetation levels almost reached those of modern times.

■ THE AUTHOR OF 'GILGAMESH' EXPERIENCED THE FEAR OF DEFORESTATION

At the end of the last-glacial period, goosefoot and mugwort plains spread across the vicinity of Ghab Valley. However, after 12,500 ^{14}C years BP, temperature and humidity rose, causing the spread of deciduous oak forests. Forests of deciduous oaks spread on the lower slopes of Mt

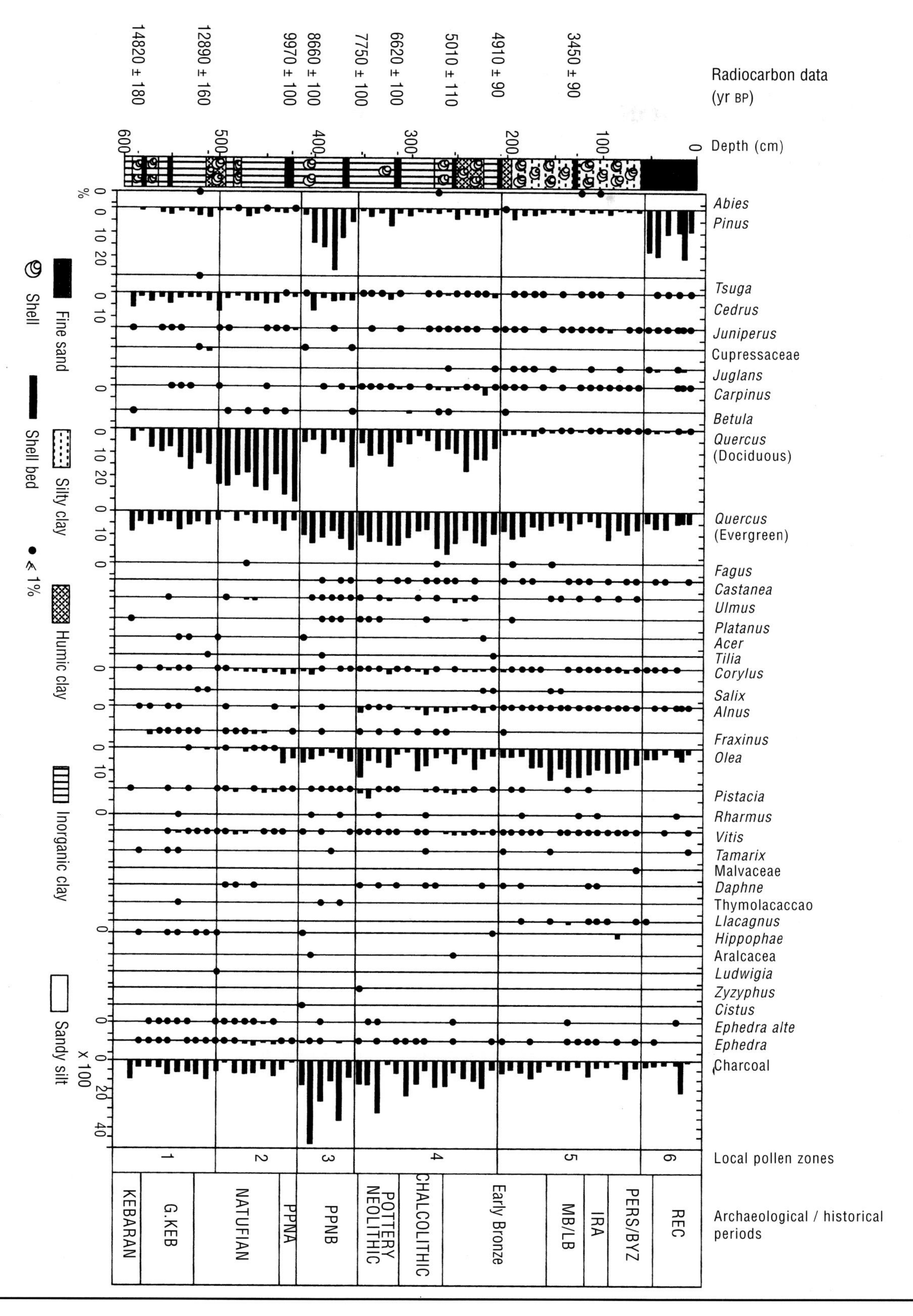

Fig. 12: Relative percentage pollen diagram from Ghab Valley, north-west Syria: arboreal pollen (AP) and charcoal (Yasuda, et al., 2000)

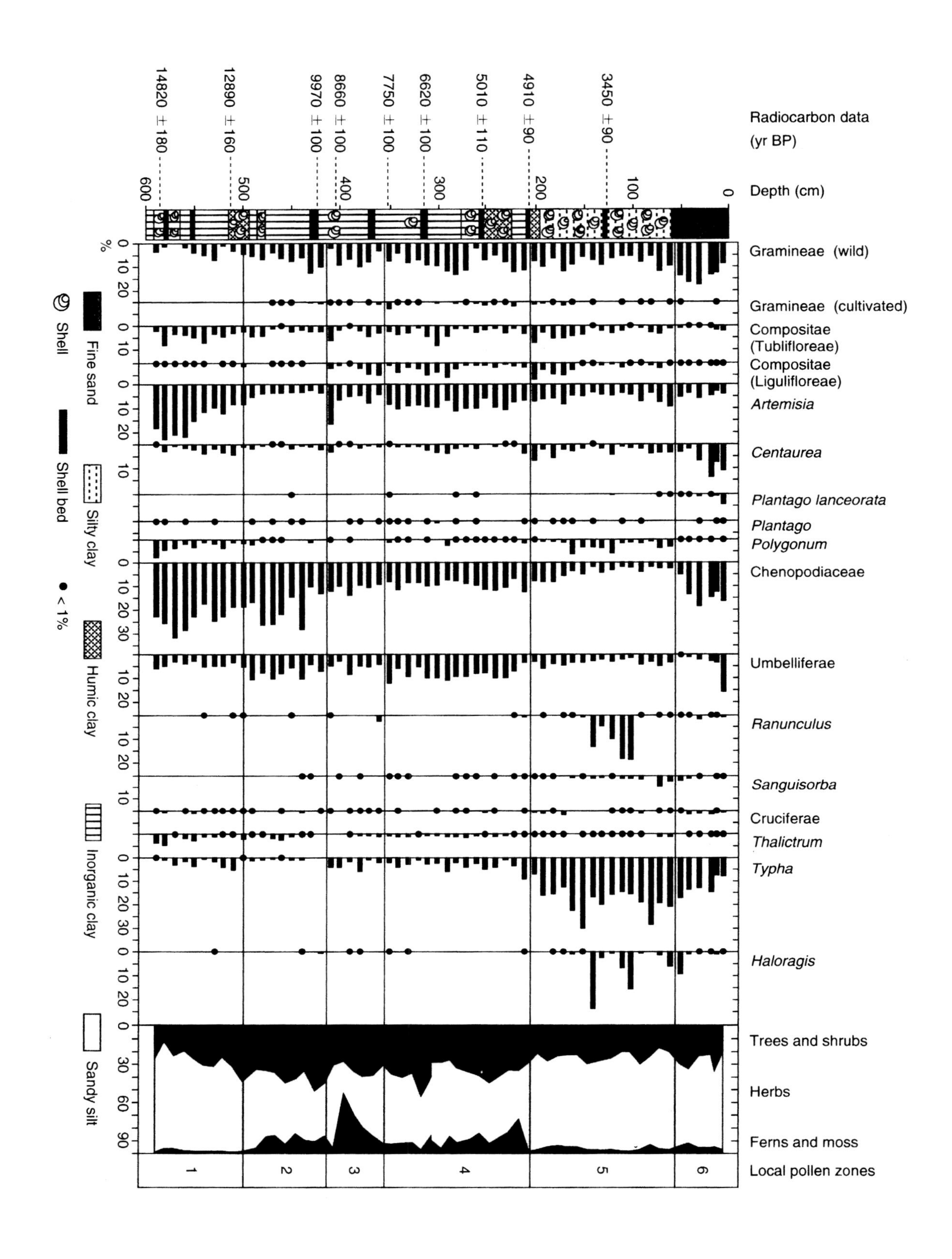

Fig. 13: Relative percentage pollen diagram from Ghab Valley, north-west Syria: non-arboreal pollen (NAP) (Yasuda, et al., 2000)

Ansarie, and forests of Lebanese cedar spread on the upper slopes.

About 9,000 ^{14}C years BP, however, deciduous oak pollen suddenly decreased, and there was a rapid increase in charcoal. In addition, pollen from secondary pine forests and olive pollen also rapidly increased. This clearly indicates the destruction of deciduous oak forests on the foothills of Mt Ansarie due to human activities of PPNB people concomitant with the cultivation of olives. Later, there was also a spread of secondary pine forests. During this period, however, Lebanese cedar, which grows at higher altitudes above sea level, had not yet incurred large-scale destruction.

At about 7,700 ^{14}C years BP, however, pine and Lebanese cedar pollen suddenly and rapidly disappear from the pollen diagram (Fig. 14). This indicates that after this period began, humans reached the Lebanese cedar, and destroyed them together with pines that had grown up as secondary forests.

In addition, from 4,900 ^{14}C years BP, deciduous oak and Lebanese cedar pollen almost ceased to appear. This period also coincides with the time of King Gilgamesh, in the early Bronze age of the Sumerian Dynasty. During the time of Gilgamesh, the massive forests of Lebanese cedar and deciduous oak completely disappeared from the eastern foothills of the Ansarie mountain range.

Contrary to my predictions, the deforestation of the eastern slopes of the Ansarie mountain range in north-western Syria had already begun at 9,000 ^{14}C years BP, and by 4,900 ^{14}C years BP, almost all the giant forests had completely vanished. From well before the time when Gilgamesh killed Humbaba, the forests of the Ansarie mountain range had already been violently destroyed. The destruction of the Lebanese cedar forests began at least 3,000 years before the killing of the forest god Humbaba.

This explains why the author of the Epic of Gilgamesh chose the destruction of the Lebanese cedar as an important motif for the poem. From well before 2600 BC when the author of the poem lived, violent deforestation had been taking place, and people already knew that the existence of forest resources was the most important issue that governed the fate of their country. Moreover, the author of the epic of Gilgamesh continued to state that the supreme God, Enlil, who knew of the murder of the forest god Humbaba, hurled down the following curse in violent anger: 'The food they eat shall be fire, and the water they drink shall be fire.' (Perlin, 1988). Enlil also wrought the divine punishment of death on Enkidu for killing Humbaba. Then, the death of his friend, led Gilgamesh to his fate of undertaking a long voyage to the netherworld. Enlil, who knew of the murder of Humbaba, devoured the land with fire and scorched all the food in his rage. This background tells us that the author of the epic was already aware that desertification would follow the destruction of the forests.

The forests around Mesopotamia had already been destroyed over 4,000 years before the epic of Gilgamesh was written. Probably, after the forests had been destroyed, the land had changed into sterile desert many times (Fig. 15). The author of the poem of Gilgamesh had no doubt already experienced or heard of the terrifying desertification that followed total deforestation.

■ KING GILGAMESH TRAVELLED AS FAR AS THE LEBANESE MOUNTAINS

The Lebanese cedar forests on the eastern slopes of the Ansarie mountain range had already been completely destroyed 4,000 years earlier, so King Gilgamesh was forced to penetrate deeper into the Lebanese mountain range. The mountain called Mt. Sariain the ancient Babylonian version, or Shirarain in the standard version of the epic of Gilgamesh, is taken today to be Mt Hermon (Fig. 15), and so undoubtedly Gilgamesh penetrated deep into the Lebanese mountain range.

Tsukimoto (1996) writes that the region to which the leaders of the early Sumerian Dynasty ventured in their expedition for the Lebanese

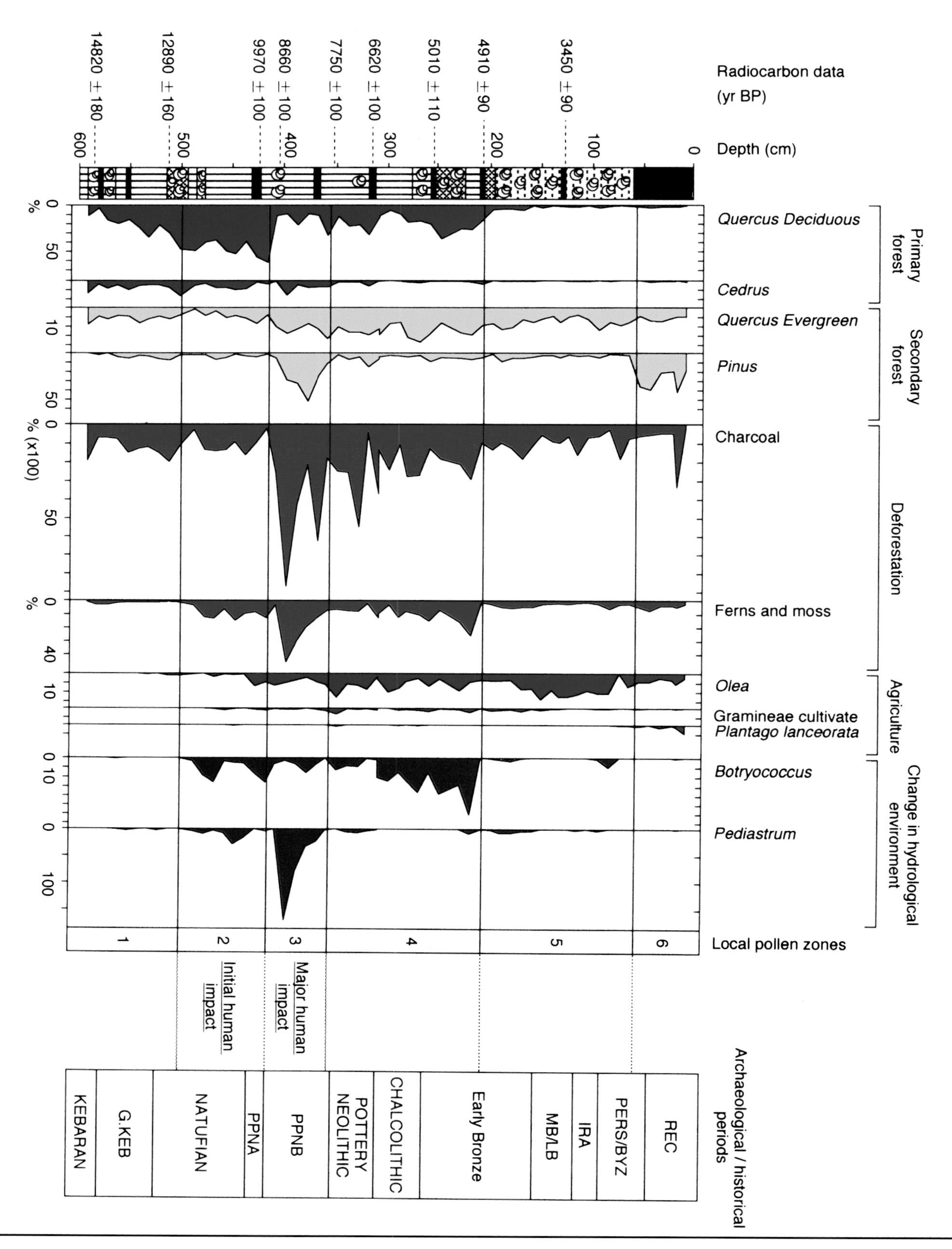

Fig. 14: Relative percentage pollen diagram for selected taxa of the Ghab Valley, demonstrating the deforestation (Yasuda, et al., 2000)

Fig. 15: Sacred Mt Hermon, covered with snow. King Gilgamesh journeyed as far as the Lebanese mountain range to the west in his search of the Lebanese cedar

Fig. 16: Only a few Lebanese cedar remained in Busshari, Lebanon (Photo by Yasuda)

cedar was the Zagros mountain range in the east, but they changed the direction to the west during the ancient Babylonian era.

For the people of ancient Mesopotamia, the east where the sun rose, the light of the rising sun itself meant the setting out on a voyage to the land of the dead. Consequently, the east symbolised a holy land, as well as the land of the dead. Doubts remain, however, as to whether this journey to the land of the dead was in fact a quest to find the forests. Tsukimoto's (1996) reasons are completely overturned by the results of our pollen analysis from Ghab Valley.

The Lebanese cedar forests on Mt Ansarie close to the Lebanese mountain range were cut down at least 3,000 years before King Gilgamesh, and at the time of Gilgamesh had already almost completely disappeared. Consequently, the people of Mesopotamia almost certainly nurtured a fierce regard for the natural resource of Lebanese cedars and their continued existence in the Lebanese mountain range behind Mt Ansarie. In all probability, it was King Gilgamesh himself who was the hero who cut down the forests of Lebanese cedar in the Lebanese mountains. The notation that 'the mountains of Saria [Hermon] and the mountains of Lebanon shook with the roars of Gilgamesh and Enkidu'[7] in the ancient Babylonian version probably tells the historical truth.

Around 2600 BC, when the Lebanese cedar forests on the Syrian plains and along the Euphrates river had been completely destroyed, Gilgamesh, the king of Uruk, penetrated the Lebanese mountains alone looking for the cedars. He then cut down the cedars in the Lebanese mountains and carried them back to the city of Uruk. It was a difficult task to transport the logs overland using several hundred head of oxen and horses to carry them from the middle of the Lebanese mountains to the Euphrates river, then transporting them by raft a further 2,000 km downstream to Uruk.

Even so, for the people of ancient Mesopotamia, who desperately needed the logs, the resources of the Lebanese cedar forests were an invaluable resource they sought. It is only natural that they should sing the praises of Gilgamesh, who accomplished this feat.

However, the people of that time already knew the fear that comes with deforestation. When they killed the god of the forest and slaughtered the Bull of Heaven, the wrath of the gods of nature and of Mother Earth fell upon them. In other words, as they destroyed nature, the people of ancient Mesopotamia were already aware that nature would retaliate with the fatal vengeance of death.

For some reason, Gilgamesh went to the land of death to meet Utnapishtim (Noah), who had survived the great flood[13]. This may imply that the great flood was caused not only by climatic change at 5,000 ^{14}C years BP, but also by the destruction of the forest.

We have to understand how the people of that time felt when they encountered the life and death experience in the midst of the destruction of nature and the drying up of natural resources described in ancient myths. Umehara (1988) has already skilfully pointed this out in his drama 'Gilgamesh' based on the world's most ancient epic poem. Passed down over 5,000 years, the epic of Gilgamesh continues to send a message to modern generations faced directly with the destruction of nature and the drying up of natural resources, and hence continued to pass on the spirit of the Lebanese cedar forest (Fig. 16).

The Spirit of the Forest in the Nihon Shoki

■ SUSANOO-NO-MIKOTO AND YAMATA-NO OROCHI (THE GOD OF STORMS AND THE GIANT SNAKE)

The *Nihon Shoki*,[14] the Chronicles of Japan, was completed in AD 720, almost 3,000 years after the epic of Gilgamesh was written. In the eight section of the first volume, Jindai, of the Urabe Kanekata version of *Nihon Shoki*,[15] the chapter

Fig. 17: A rare picture of Yamata-no-orochi drawn by foreign scholars studying Japanese culture at the beginning of the twentieth century (James and Globle, 1910)

where the jewelled sword appears, the tale of Susanoo-no-mikoto exterminating Yamata-no-orochi is repeated four times in the main text and part of the annotations, while the content of the tale changes for some reason. Let us compare the content of the tale of Yamata-no-orochi's extermination, repeated four times, in the order in which they appear.

First, the tale of Yamata-no-orochi's extermination that opens the main text has the best and most detailed literary expression. When Susanoo-no-mikoto descends from heaven to the Hii river in Izumo-no-kuni (modern-day Shimane Prefecture), he hears the cries of an elderly couple coming from upstream. When Susanoo-no-mikoto asks them why they are crying, the elderly couple who are local deities called Ashinazuchi and Tenazuchi, explain that they used to have eight daughters, but every year one had been swallowed by Yamata-no-orochi, and now the last daughter, Princess Kushinada, is going to be swallowed, and there is no way of saving her. Susanoo-no-mikoto asks whether, if he was to save her, would they give him the girl. The elderly couple's response is that they would willingly do so.

Susanoo-no-mikoto instructs Ashinazuchi and Tenazuchi to fill eight containers with sake and wait for Yamata-no-orochi (Fig. 17) to come. Yamata-no-orochi has eight heads, his eyes are

Fig. 18: Susanoo-no-mikoto (Yaegaki-jinjya, Shimane Prefecture, Japan, Umehara, 1980)

red, trees are planted on his back, and he is so large that he spans eight hills and eight valleys. When Yamata-no-orochi finds the sake barrels, he dips a head into each barrel, becomes drunk, and falls asleep. In that instant, Susanoo-no-mikoto cuts Yamata-no-orochi to ribbons with his broadsword. However, the sword fails to cut Yamata-no-orochi's tails. Thinking this odd, when Susanoo-no-mikoto tries to slice apart the tails, a new broadsword falls out. This is the so-called Sword of Kusanagi. Susanoo-no-mikoto declares that this is a divine sword, and that he cannot take it, so he offers it to goddesses in heaven. Then he builds a shrine at Suga in Izumo and marries Princess Kushinada, and they have a baby, Oanamuchi-no-kami. Susanoo-no-mikoto then declares Ashinazuchi and Tenazuchi as the heads of his child's shrine.

The second story (in Vol.1, the second appendix) and third story (Vol. 1, the third appendix) are almost the same as the first story. However, when we come to the fourth story (Vol.1, the fourth appendix) its content is completely different. Susanoo-no-mikoto takes the child, Itakeru, and descends from heaven at Silla, Korea. However, because this country does not need them, they depart for Izumo (modern Shimane Prefecture) by boat. Then, when they arrive upstream of the Hii river, the giant man-eating snake is waiting. Susanoo-no-mikoto exterminates the giant snake with his sword, Amano-hahagiri and when he cuts off its tail, a broadsword falls out. This is the so-called Sword of Kusanagi.

In this fourth story, none of the characters of the earlier stories, Ashinazuchi, Tenazuchi, or Princess Kushinada, appear. This is simply a story of Susanoo-no-mikoto exterminating the giant snake with the sword Amano-hahagiri. Here, 'Haha' means 'Giant snake.' Thus, the gods of heaven use this sword to kill a giant snake that symbolises the gods on earth.

■ GILGAMESH AND SUSANOO-NO-MIKOTO

When we compare the tale of *Nihon Shoki* with the ancient Mesopotamian epic of Gilgamesh, there are obvious similarities. The tale of Susanoo-no-mikoto (Fig. 18) exterminating Yamato-no-orochi is very similar to that of Gilgamesh slaying Humbaba. Gods in human form with new metal weapons such as a bronze axe or broadsword exterminate Humbaba and Yamato-no-orochi as manifestations of nature which came first. Susanoo-no-mikoto and Gilgamesh fulfil the same role. Gilgamesh is a symbol of the start of a civilisation, which has new weapons made of bronze, and Susanoo-no-mikoto is a symbol of the start of an era of new weapons made of iron (Yasuda, 1994).

These two myths from east and west are tales of the battle between the old gods and the new ones, the old in the form of the monstrous

Humbaba, who lives in the forest, and the giant snake. Pines and oak trees are planted on Yamata-no-orochi's back, hinting at a fundamental connection between the snake and the forest. Humbaba and Yamata-no-orochi are gods of the forest. The gods with the new metal weapons, Gilgamesh and Susanoo-no-mikoto are male ones in human form.

There are however distinctive differences between the two myths from east and west. This is noted in the latter part of the fourth story of the legend of Yamata-no-orochi, which can be viewed as closest to the original prototype. An extract (Vol.1, the fourth and fifth appendices) is quoted below:

> Before this, when Susanoo-no-mikoto descended from Heaven, he took down with his son Itakeru the seeds of trees in great quantity. Susanoo-no-mikoto said: There are many gold and silver treasures in Silla [Korea]. However, there were no such treasures in Japan. When asked to provide other treasures, therefore, he plucked hairs from his beard and where he threw them, they became Japanese cedars [*Cryptomeria japonica*] (Fig. 19), the hairs from his chest became cypresses [*Chamaecyparis obtusa*], the hairs from his hip became umbrella pine [*Sciadopitys verticillata*], and the hairs from his eyebrows became camphors [*Cinnamomum camphora*]. Then he said the following: he will make the Japanese cedar and the camphor into boats, the cypress he will use to build a palace, and the umbrella pine he will use to fashion a coffin. Furthermore, when his son Itakeru, descended from heaven, he brought with him the seeds for many different trees and spread them all throughout the land, starting with Tsukushi [modern-day Kyushu], and created verdant mountains.Thus green mountains were produced. For this reason Iso-takeru no Mikoto was styled Isaoshi no Kami [the meritorious God]. He is the Great Deity who dwells in the Land of Kii [modern-day Wakayama Prefecture][16].

Cypress was the best wood for construction purposes, and people were already using umbrella pines to make the majority of sarcophagi in the Yayoi Period, so the tradition of using trees recorded in the *Nihon Shoki* can be understood to date from the Yayoi Period onwards. Besides, when Susanoo-no-mikoto's son, Itakeru, descended from heaven, he brought with him the seeds for many different trees and plants. He did not, however, plant them in Silla, but brought them all back to Japan with him, and spread them throughout the land, starting with Tsukushi, and created verdant mountains.

There is thus a crucial difference between Gilgamesh and Susanoo-no-mikoto. After killing the god of the forest, Humbaba, Gilgamesh and Enkidu are able to freely take the Lebanese cedars, which, it is noted, they cut down five times, and transport them along the Euphrates river by raft to Nippur and Uruk. However, not only does Susanoo-no-mikoto, who exterminates Yamata-no-orochi, extract his own hair to grow useful trees such as cedar, cypress, camphor, and umbrella pine, but his son Itakeru scatters seeds of plants throughout Japan and creates forests.

This is clearly indicative of different points of view of nature in the myths from east and west, and their approach to forests. The large tracts of forest along the Mediterranean coast from Mesopotamia were destroyed completely, exactly as the myth relates. The people who came later also completely destroyed the forests in the same way as Gilgamesh did.

What then is the kind of vegetation change that can be identified in the Japanese archipelago, where the gods scattered seeds that gave rise to the trees? Did the gods of Japan continually create verdant mountains just as the myth relates? Let us examine the changes that occurred in the Japanese archipelago.

■ RESULTS OF POLLEN ANALYSIS IN THE JAPANESE ARCHIPELAGO

Results of pollen analysis from Daira-ike mire, Shiga Prefecture:

The Daira-ike mire is located at 490 m above sea level (35° 27' N., 135° 50' E.), on the north-west

Fig. 19: Japanese cedar (*Cryptomeria japonica*), Haguro mountain in Yamagata, Japan (Photo by Yasuda)

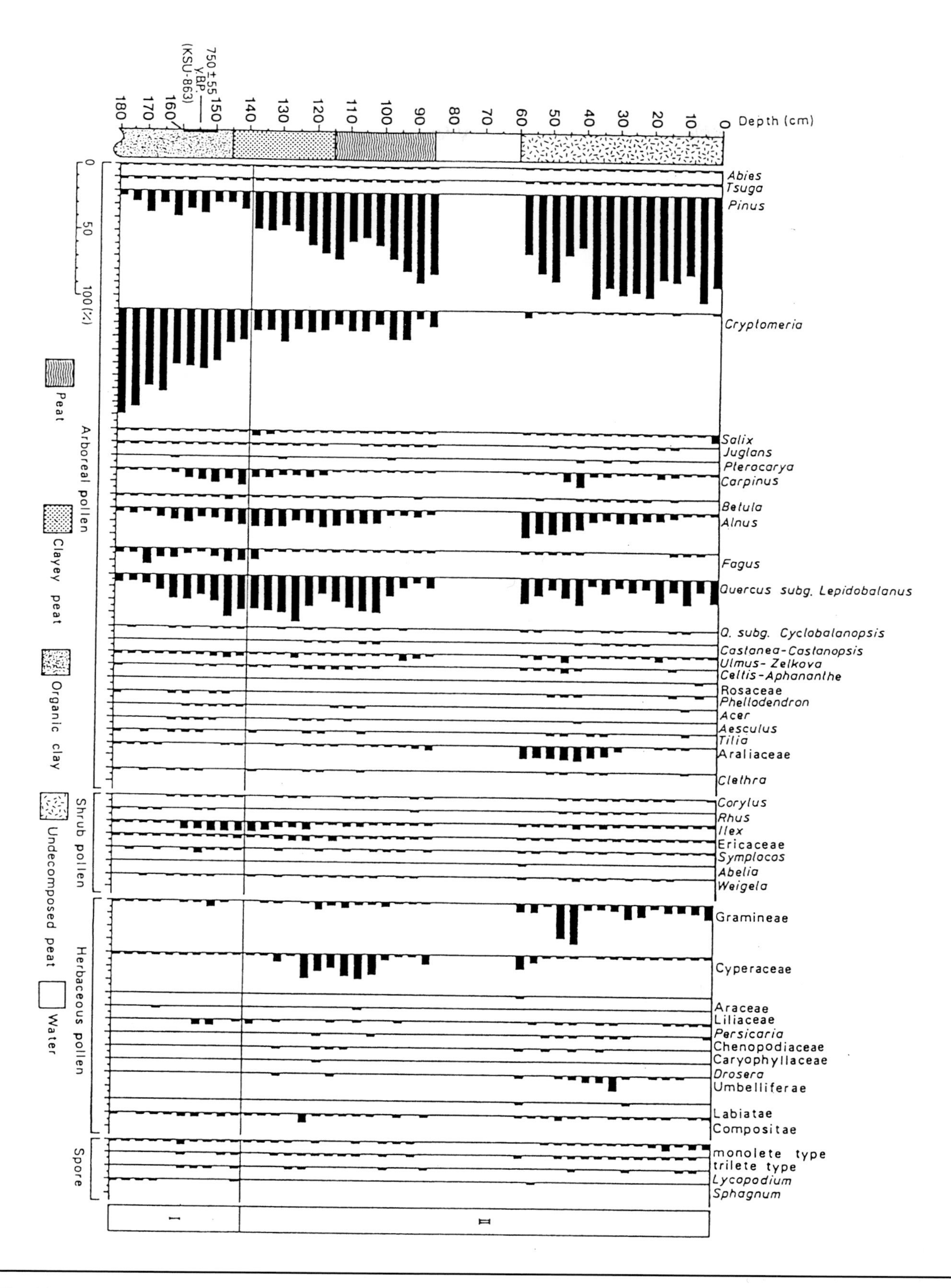

Fig. 20: Relative percentage arboreal pollen diagram of the Daira-ike mire, Shiga Prefecture Japan (Yamaguchi, et al., 1989)

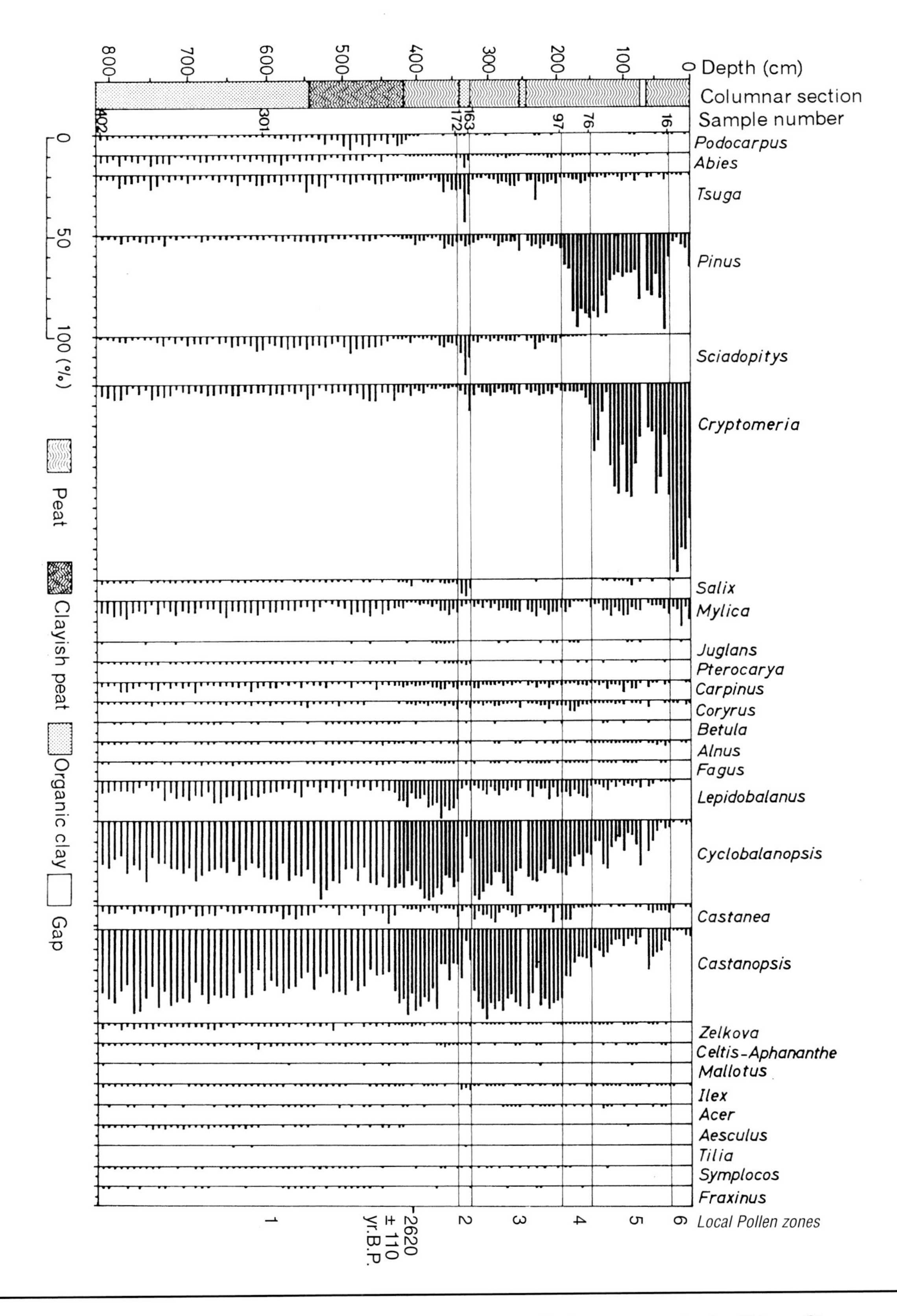

Fig. 21: Relative percentage arboreal pollen diagram from Ukishima-no-mori mire, Shingu City, Wakayama Prefecture (Takeoka et al., 1983)

coast of Lake Biwa in Shiga Prefecture. Fig. 20 shows the results of the pollen analysis from the Daira-ike mire (Yamaguchi, et al., 1989). From 15 cm below the horizon yielding 750 ± 55 ^{14}C years BP, the Japanese cedar (*Cryptomeria*) pollen reduce from 70% to approximately 30%. In addition, from the upper 15 cm of the horizon in 750 years BP, the pollen reduce to 20% at the boundary between local pollen assemblage zones I and II, and also in the same horizon, there is a dramatic reduction in beech trees (*Fagus*). The Japanese cedars then disappear in the horizon at a depth of 85 cm. Whereas the Japanese cedars unilaterally reduce here, there is an increase in a deciduous oak (*Quercus Lepidobalanopsis*), alders (*Alnus*) and, some time later, pines (*Pinus*).

The results of the pollen analysis show that a 1,000 years ago, the Japanese cedars around the Daira-ike mire had been cut down, and that following the deforestation, secondary forests of deciduous oak, alders, and Japanese red pines had expanded. A long time ago Japanese cedars and Hinoki cypress were cut down in the region close to Kyoto (the ancient capital city of Japan) for use as building material.

Results of Pollen Analysis from Ukishima-no-mori mire, Shingu City:
The *Nihon Shoki* recounts how Itakeru was venerated in Kii-no-kuni (modern-day Wakayama Prefecture) as a powerful god because of his success in making Japan's mountains lush and verdant by scattering seeds of trees. Fig. 21 shows the results of the pollen analysis from the Ukishima-no-mori mire, Shingu City, Wakayama Prefecture (33° 43' N., 136° E., 7 m above sea level) (Takeoka, et al., 1983). This pollen diagram is classified into local pollen assemblage zones 1 to 6 from the bottom upwards. There are major changes between local pollen assemblage zone 3 and 4, approximately two metres above the horizon yielding 2620 ± 10 ^{14}C years BP. First, there is a reduction in the major tree pollen which form laurel forests such as the evergreen oak (*Quercus Cyclobalanopsis*), chinquapin (*Castanopsis*), bayberries (*Myrica*), and instead there is an increase in pines (*Pinus*). This clearly speaks of the destruction of the laurel forests such as evergreen oak and chinquapin due to the deforestation by human activities, and instead the expansion of pines as secondary forestation.

From calculations based on the hypothesis that accumulation of detritus took place at a constant rate, and from ^{14}C dating it would seem that this forest destruction took place approximately 1,300 years ago. Around 1,300 years ago, in other words, in the latter half of AD seventh century, the laurel forests largely comprising evergreen oaks and chinquapin of Ukishima-no-mori mire on the southern Kii Peninsula were destroyed, causing an expansion of secondary forest such as red pines.

However, approximately 300 years after that, there is a second major change between local pollen assemblage zones 4 and 5. Japanese cedar pollen increased unilaterally after this time, carving out the landscape of cedar-dominated forests we know today.

This Japanese cedar increase, concomitant with the start of local pollen assemblage zone 5, resulted from human intervention in the forests, namely, humans planting trees. Ancient people began planting cedars in the vast tracts of land covered by secondary red pine forest, following the destruction of the primeval laurel forests. This happened approximately 1,000 years ago. From the results of the pollen analysis, tree planting had already begun in the Heian Period. This increase in the upper part of the pollen diagram for Japanese cedar pollen is an identifiable phenomenon not only in the Ukishima-no-mori mire, but generally all over the Japanese archipelago from southern Honshu onwards (Yasuda, 1991).

As with the example of the Daira-ike mire in Shiga Prefecture, whereas the Japanese cedar teees were destroyed, only in extremely rare cases did they not grow again. The topsoil of the area around the Daira-ike mire was washed away due to this deforestation, resulting in an environment in which only secondary forests of red pine could grow.

▪ THE AUTHORS OF THE *NIHON SHOKI* WERE AWARE OF THE IMPORTANCE OF NATURAL FOREST RESOURCES

As can be presumed from the results of the pollen analysis from the Daira-ike mire and the Ukishima-no-mori mire, around the eighth century when the *Nihon Shoki* was compiled, the Japanese cedar and Hinoki cypress trees around the capital of Nara appear to have been considerably deforested for use as building materials. The authors of the *Nihon Shoki* were undoubtedly sufficiently aware of the importance of natural forest resources. Tokoro (1980) has already pointed out that from around the time of the Taika Reform in AD seventh century, the forests were laid waste to build the capital city, and it was in all probability necessary to take urgent political action to protect the mountain forests. This phenomenon was limited to the vicinity of the capital city, and corresponds well with the results of the pollen analysis. The legend in the *Nihon Shoki* that Susanoo-no-mikoto gave birth to trees, and that Itakeru-no-mikoto scattered seeds of trees was probably born out of an increase in demand for natural forest resources and the drying up of those resources.

There are written documents relating to tree planting, dating from AD 1364 and endorsed by the Tada Shrine in Kawanishi City, Hyogo Prefecture. These read, 'To reverently state to use trees for cooking in the kitchen and for glazing tiles for the main temple hall, so to grow trees in recent years,' which means there is a record of 'planting' trees as material for cooking and glazing tiles. This record of 'planting' can be taken as the oldest written one of its kind, as Murai (1987) has pointed out.

From these ancient documents, we can see that tree planting had definitely started as early as the fourteenth century in Japan. The results of the pollen analysis indicate a high possibility that planting actually started long before that. Just as with Susanoo-no-mikoto, who plucked out his beard and scattered it to create Japanese cedar, and Itakeru-no-mikoto who scattered tree seeds throughout Japan, the Japanese began planting trees long ago. In all probability, people were aware of the importance of tree planting in the eighth century when the *Nihon Shoki* was completed, and the authors of both the ancient chronicles *Kojiki* and *Nihon Shoki* were aware of the importance of planting trees on the mountains.

Similarly, the author of the epic of Gilgamesh was also aware of the importance of the forests, and understood the fear of their destruction. He was also aware that if there were no forests, the life-giving water would dry up and desertification would occur. In Mesopotamia, located in a far dryer region than Japan, such damage would be far graver than in Japan. It would be on a scale that could throttle the civilisation. However, unlike the authors of the *Nihon Shoki*, he never thought of planting trees himself.

Tomiyama (1987) has taken a paragraph attributed to Susanoo-no-mikoto and has pointed out that in Japan, culture was built upon the cultivation of forests, whereas other civilisations of the world were built upon their destruction. About this, Suzuki (1993) has commented that, 'The myths of planting sacred trees can be seen in mythology throughout the world. Just because the myths tell of sowing tree seeds, it is rash to read this directly as ideas of forestation, and to take it as something unique to Japan.' It is not true, however, that similar tales of tree planting appear in myths throughout the world. This is clear from the extract of the epic of Gilgamesh that we have quoted. We ought to examine the myths of the

Fig. 22: Forests were consumed by sheep and goats, Sultansazul Göl in Turkey (Photo by Yasuda)

world a little more closely. We can definitely conclude from a single quote by Susanoo-no-mikoto that the forestation notion, i.e. harmonious coexistence with nature, existed at a fundamental level in Japanese culture. It is reasonable to take the view that it tells of an affinity with the forest as being a special Japanese characteristic, as Tomiyama (1987) has pointed out.

Comparative Study of the Myths and History of Forest In East and West Asia

So why is it that, while appreciating the importance of the forests, the authors of the epic of Gilgamesh did not create myths about planting trees? One key to unlocking this mystery is the existence of the Bull of Heaven. After slaying the forest god Humbaba, Gilgamesh fought with the Bull of Heaven (see Fig. 8). The Bull was a symbol of Mother Earth, representing Mesopotamia together with Humbaba.

The form of cultivation in ancient Mesopotamia and the world along the Mediterranean coast was a fusion of wheat cultivation and livestock farming. The Bull of Heaven represented oxen, sheep, and goats as livestock. After clearing the forests to create farmland, people put their livestock out to pasture. Livestock provided not only milk and a source of protein, but also wool and fur. Seen from this perspective, there was more benefit to be gained not from encouraging the forests to grow, but to clear and use them, and then set the livestock to graze on the land. (Fig. 22).

Fig. 23: *Satoyama*, secondary deciduous oak forest in Ono City, Fukui Prefecture (Photo by Yasuda)

Consequently, the notion of planting trees to regenerate forests probably did not arise for a long time among the people living in the land on which wheat was cultivated and livestock farmed. This is clear if we look at the fact that in Europe, where wheat cultivation and livestock farming came as one set, tree planting did not occur until modern times. Thus, forests from Mesopotamia to the Mediterranean and further into Europe were ruthlessly destroyed, and there was no attempt at reforestation until very recently (Fig. 22).

In contrast, the Japanese, because of their culture of rice cultivation, used fish as a source of protein rather than livestock. Although livestock for consumption could be found in China and on the Korean Peninsula, it did not spread to Japanese society. Certainly, there is the discovery of pigs from the Yoshinogari site of AD second century in Saga Prefecture, but this breeding disappeared at some point in time.

Why did Japan's Yayoi people refuse to eat livestock? I believe the background to this is the cultural tradition of forests that had existed for so long by then in the Jomon period (Yasuda, 1996). Warmth and high precipitation in the Japanese archipelago following the destruction of the primeval laurel forests resulted in the regeneration of secondary forest such as deciduous oaks (*Quercus serrata*, *Q. acutisima*) and red pines. As we have already seen, the results of the pollen analysis from the Daira-ike mire (Fig. 20) show this clearly. The secondary forests with a *'Satoyama'* culture, which has a special life connection with the forests, was the core of the farming structure created for Japanese rice cultivation for which they were essential. It was

Fig. 24: *Satoyama*, secondary red pine forest and rice paddy field in Tsuruga City, Fukui Prefecture (Photo by Yasuda)

the '*Satoyama*,' (Figs. 23, 24) secondary forests that provided the undergrowth used to cultivate and fertilise the water of the paddy fields.

Moreover, all construction in Japan, including that of the imperial palace, relied on wood. This was very different from Mesopotamia, where sun-baked bricks and stones were used to build its palaces and temples. Consequently, the ruling statesmen also woke up quickly to the importance of regenerating natural forest resources.

However, it appears that the degree to which the ruling statesmen woke up to that importance was clearly higher in Mesopotamia. The reason is that in Mesopotamia, which has no forests, they had to import lumber even from far upstream of the Indus River. In spite of this, tree planting did not spread throughout Mesopotamia or along the Mediterranean coast.

In Japan, which is blessed with forestry resources, tree planting began at an early stage, whereas in Mesopotamia, which lacks such resources, tree planting and cultivation never happened. This testifies that forests were seen as an important occupational area by the Japanese. People considered their work as part of a set, including rice cultivation and fishing, against a background of a climate and topography with high precipitation and warmth, which invigorated the regeneration of the trees. We should view this as an expression of their special relationship with forests. People became aware at a very early stage of the importance of the forests and of planting trees, and so they set about planting trees on mountains as an important task to be performed and this is a unique aspect of Japanese culture.

Fig. 25: Japanese Shinto shrine and the holy forest (Chinjyu-no-mori) relate the tradition of the Japanese forest civilisation, Oomiwa Jinjya, Nara Prefecture (Photo by Yasuda)

In the *Nihon Shoki*, Susanoo-no-mikoto is quoted as saying, 'Cedar and camphor; these two trees should be made into the floating treasure (i.e. ships), Hinoki cypress is to be used as timber for building fair palaces, umbrella pine is to form coffins.'[16] Just as is indicated by the quotation Japanese venerated the forests as treasures for 1,200 years (Fig. 25).

Conclusion

The two myths from the east and the west tell of the existence of a god who cut down the forests and used them to bring prosperity to the people, and of a god who raised the forests and gave birth to the trees. Under the former god, the forests and civilisation were completely destroyed. Under the latter, the forests are protected and civilisation thus maintained. In our present situation, where we are directly faced with world deforestation and a global environmental crisis, it goes without saying that we must adhere to the tenets of the latter god who protects the forests.

In addition, environmental archaeological research, supported by modern technology, has proven that the ancient myths recount some measure of the truth. It has also become possible to read the spirit of the forests, which the authors of the myths wanted to pass on. The author of the epic of Gilgamesh was already afraid of deforestation 4,600 years ago, whereas the authors of the *Nihon Shoki* knew that planting and protecting forests was a lifeline for the island nation of Japan. Today, at the beginning of the twenty-first century, it is essential that modern people, directly faced with a global environmental crisis, learn to read and understand the spirit of the forest from these ancient myths which have begun to recount a completely new system of knowledge.

Notes

1. Excerpts from the epic of Gilgamesh cited in this chapter come from the online version of the epic.
2. The Old Babylonian version. Y. Tablet 3. 5f, 7, 8.
3. Op. cit., Y. Tablet 3. 16, 17, 18f, 20.
4. Op. cit., Y. Tablet 5. 1, 3, 6, 7.
5. Op. cit., l. 23'-24', 25', 26', 27'.
6. Op. cit., l. 8'.
7. Op. cit., l. 34'.
8. The Standard version. 5. Tablet 1. 1, 2, 3, 4, 5, 6, 7, 8.
9. Op. cit., 6. Tablet 5. 4, 7, 8, 10.
10. Op. cit., 8. Tablet 1. 1.
11. Op. cit., 8. Tablet 2. 10, 11, 12, 13, 14, 15, 16.
12. Op. cit., 8. Tablet 3. 6, 7.
13. The Old Babylonian version. H. Fragments 3. Tablet 4. 10–11.
14. *Nihongi: Chronicles of Japan from the Earliest Times to* AD *697*, trans. W. G. Aston. (Tokyo: C. E. Tuttle, 1972).
15. See the eighth paragraph in the section of Jindai in Vol. 1, the first, second, third and fourth appendices in *Nihongi*.
16. *Nihongi*, p.58.

References

Baruch, U., S. Bottema (1999): A New Pollen Diagram from Lake Hula, in H. Kawagoe, G. W. Coulter and A. C. Roosevelt (eds.): *Ancient Lakes: Their Cultural and Biological Diversity*, Kenobi Production, Belgium, pp. 75–86.

Murai, Y. (1987): Introduction, in *Inagawa-cho shi*. vol.1, 1 ed. Inagawa-cho shi henshu senmon iinkai, pp. 6–10.

Niklewski, J. and W. Van Zeist, (1970): A Late Quaternary Pollen Diagram from Northwest Syria, *Acta Botanica Neerlandica*, 19: 737–54.

Perlin, J. (1988): *A Forest Journey: The Role of Wood in the Development of Civilization*. W. W. Norton & Company, New York, 445 pp.

Suzuki, S. (1993): The Study of the Conservation of Forest. *Nihonjin to Nihonbunka no Keisei*, ed. K. Hanihara, Asakura Shoten, Tokyo, pp. 171–89.

Takeoka, S. and H. Takahara, (1983): Changes of the Forests Around the Ukishima-no-mori moor at Shingu, Wakayama Prefecture, based on pollen analysis, *Nichirinron*, 94: 383–85.

Tokoro, M. (1980): The Study of the Early Modern Forestry (*Kinse Ringyoshi no Kenkyu*). Yoshikawa Kobunkan, Tokyo, 450 pp.

Tomiyama, K. (1987): Water Journey: Rediscovery of Japan (*Mizuno tabi : Nihon saihakken*), Bungei Shunju-sha, Tokyo, 235 pp.

Umehara, T. (1980): *Kojiki*, Gakushukenkyu-sha, Tokyo, 198 pp.

—— (1988): Gilgamesh (*Girugameshu*), Shincho-sha, Tokyo, 235 pp.

Van Zeist, W. and H. Woldring, (1980): Holocene Vegetation and Climate of North-western Syria. *Palaeohistoria,* 22: 111–25.
Yamaguchi, K. and H. Takahara, S. Takeoka, (1989): Forest Changes in the Lower Mountains North-west of Lake Biwa Since About 1,000 y, B.P. *Bulletin of the Kyoto Prefectural University Forests*, 33, 1–6.
Yasuda, Y. (1990): Climatic Change and the Rise and Fall of Civilizations (*Kiko to Bunmei no seisui*), Asakurashoten, Tokyo, 358 pp.
—— (1991a): Climatic Changes at 5000 years BP and the Birth of Ancient Civilizations. *Bulletin of the Middle Eastern Culture Center in Japan*, IV: 203–18.
—— (1991b): Japanese and Japanese Cedar, *Bulletin of the International Research Center for Japanese Studies*, 4: 41–112.
—— (1993): Climate Transform the Civilizations (*Kikou ga bunmei o kaeru*), Iwanami-shoten, Tokyo, 1993, 116 pp.
—— (1994): Snake and Cross: Climate and Religion in East and West (*Hebi to Jujika:Tozai no fudo to shukyo*), Jinbun shoin, Kyoto, 1994, 237 pp.
—— (1996): Forest Culture in Japan: From Jomon to Future (*Mori no Nihonbunka: Jomon kara Mirai e*), Shin Shisakusha, Tokyo, 233 pp.
Yasuda, Y., H. Kitagawa and T. Nakagawa, (2000): The Earliest Record of Major Anthropogenic Deforestation in Ghab Valley, North-west Syria: A Palynological Study. *Quaternary International*, 73/74: 127-136.

CHAPTER 2

Changes in the Forest Environment of Western Europe and the Rise and Fall of Civilisation: A Case Study in France

J.-L. DE BEAULIEU, M. BARBERO, M. REILLE,
H. RICHARD, AND D. MARGUERIE

Introduction

At the end of a century marked by the triumph of industrial civilisation, Man now has at his disposal technology that confers on him almost god-like powers. However, just as this status has been achieved, there are dire warnings that the environment has been severely damaged by human action over the past century and a half: biodiversity has diminished, unsustainable resources are being exhausted, and there is a risk of serious climate change.

This chapter examines the impact of prehistoric and historic civilisations on forest ecosystems, with a particular focus on France.

History of West European Forests

QUATERNARY CYCLES

How does the vegetation dynamics of periods when the natural environment was not affected by human impact compare with that of the present interglacial which has been deeply marked by human action? Recently our team obtained and described a very rare long pollen sequence spanning the Middle and Late Pleistocene (Reille and de Beaulieu, 1995) which allowed such a comparison.

Pollen data was obtained from cores retrieved from paleo-lake beds situated on the basalt plateau of Velay in the Massif Central (Fig. 1). A composite pollen sequence (Fig. 2) shows that glacial phases alternated with temperate phases characterised by interglacial vegetation. An absolute Ar/Ar dating of trachytic tephra, and correlation with deep-sea and continental data, enabled us to assign an age of 400,000 years to the bottom of the sequence.

The main interglacials recorded in Fig. 3 are compared in order to illustrate the natural variability of vegetation cycles. Note that each interglacial shows the same succession of forest types, namely:

1. a pioneer vegetation with pine and birch;
2. a vegetation with deciduous and more or less thermophilous forests;
3. a vegetation with conifers that today are mountain trees;
4. a vegetation consisting of boreal forests with pine and birch.

However, each interglacial had its own characteristics in regard to the relative abundance of tree species. In the lower interglacial (the Pracaux Interglacial), fir and beech show major expansion but hornbeam is only modestly represented. In the Landos

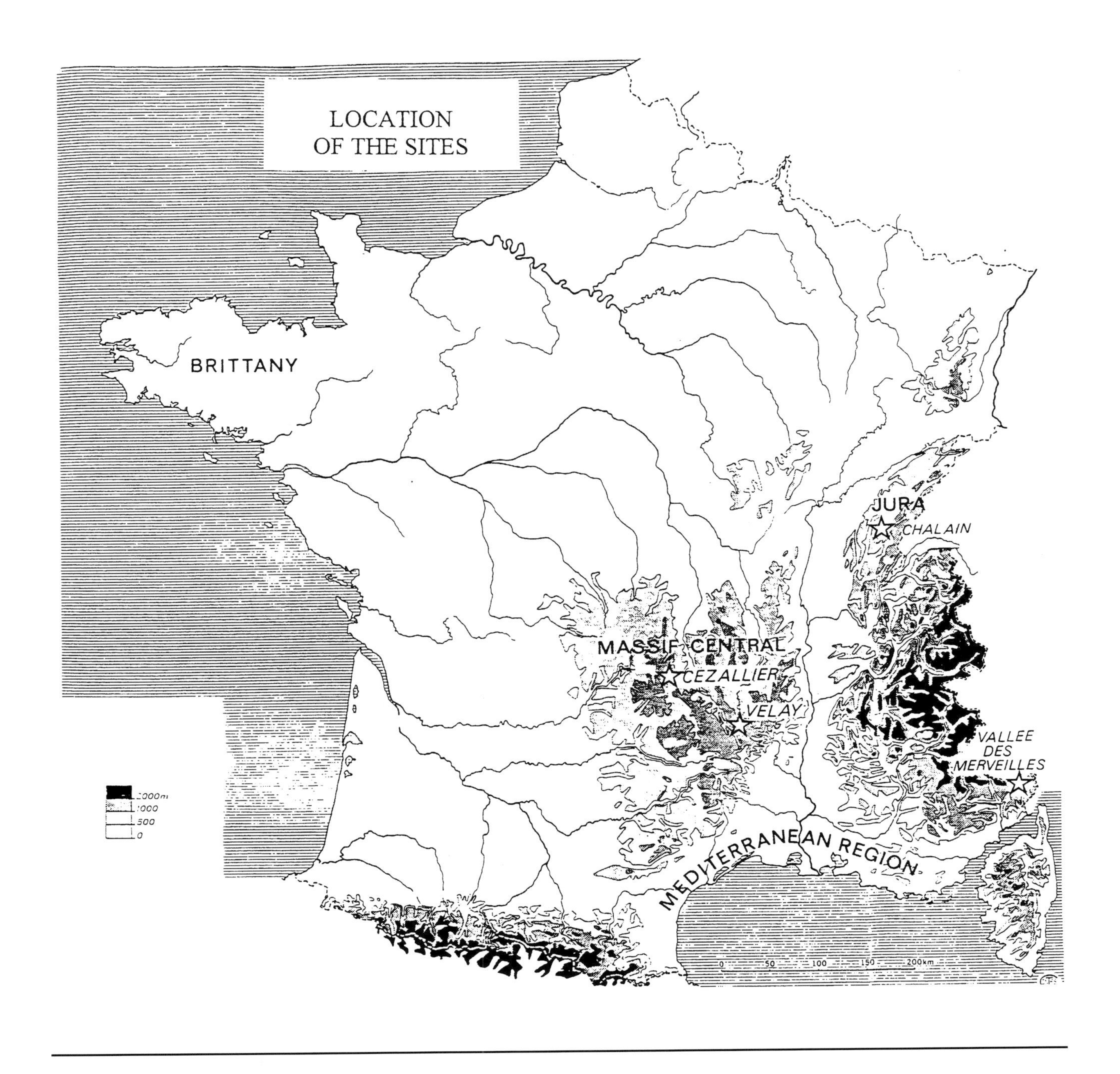

Fig. 1: Location of the sites referred to in this paper

Interglacial hornbeam is abundant, and in the Bouchet Interglacial spruce and fir play a very modest role. In the Ribains Interglacial beech is almost totally absent but in the present Holocene Interglacial beech is very abundant, whereas neither hornbeam nor spruce are common. These differences can be explained by two factors:

1. competition between plants from the beginning of the interglacial;
2. the climatic conditions during each interglacial which, according to the Milhankovich theory (Berger, 1977), are related to astronomical forcing.

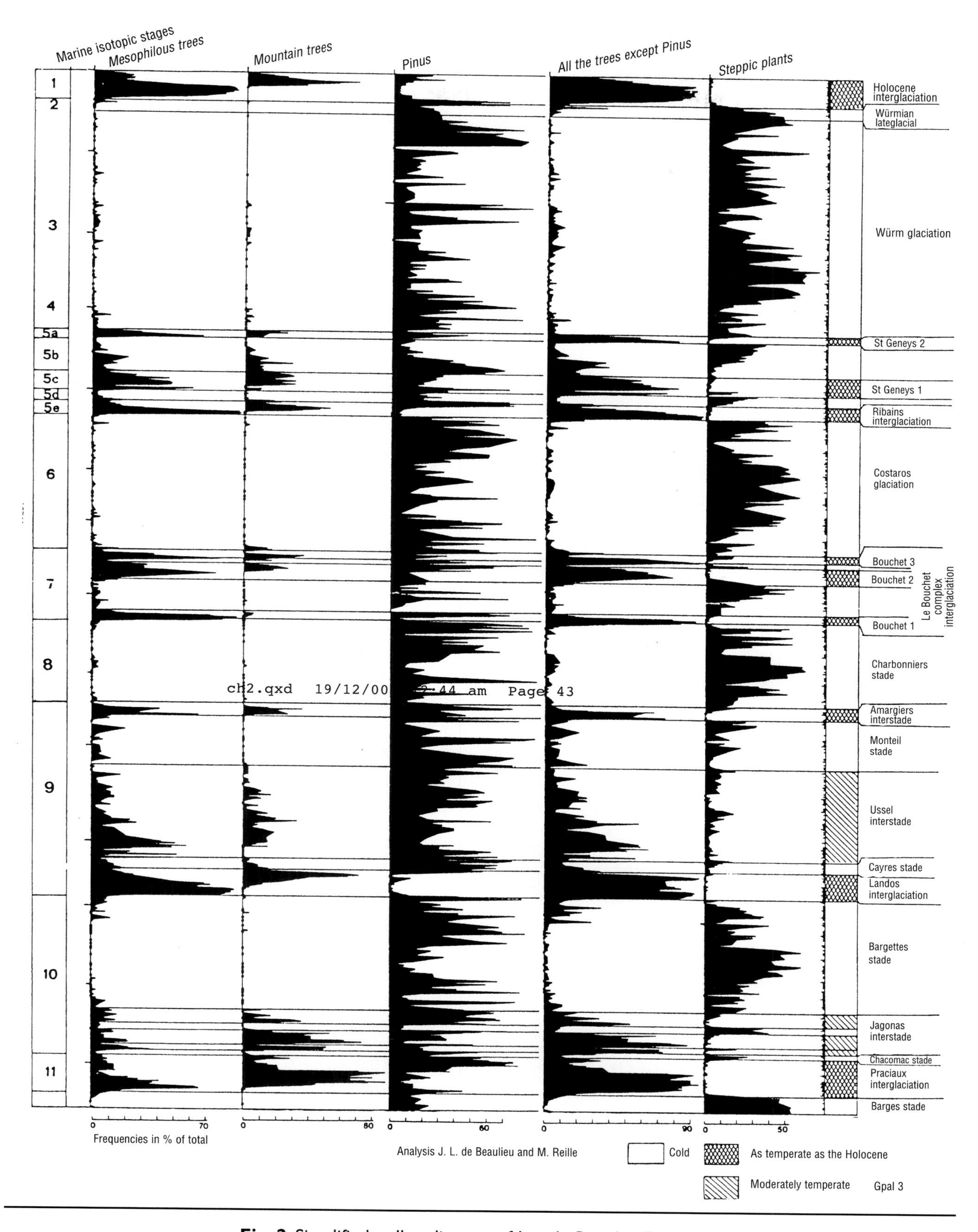

Fig. 2: Simplified pollen diagram of Lac du Bouchet/Praclaux

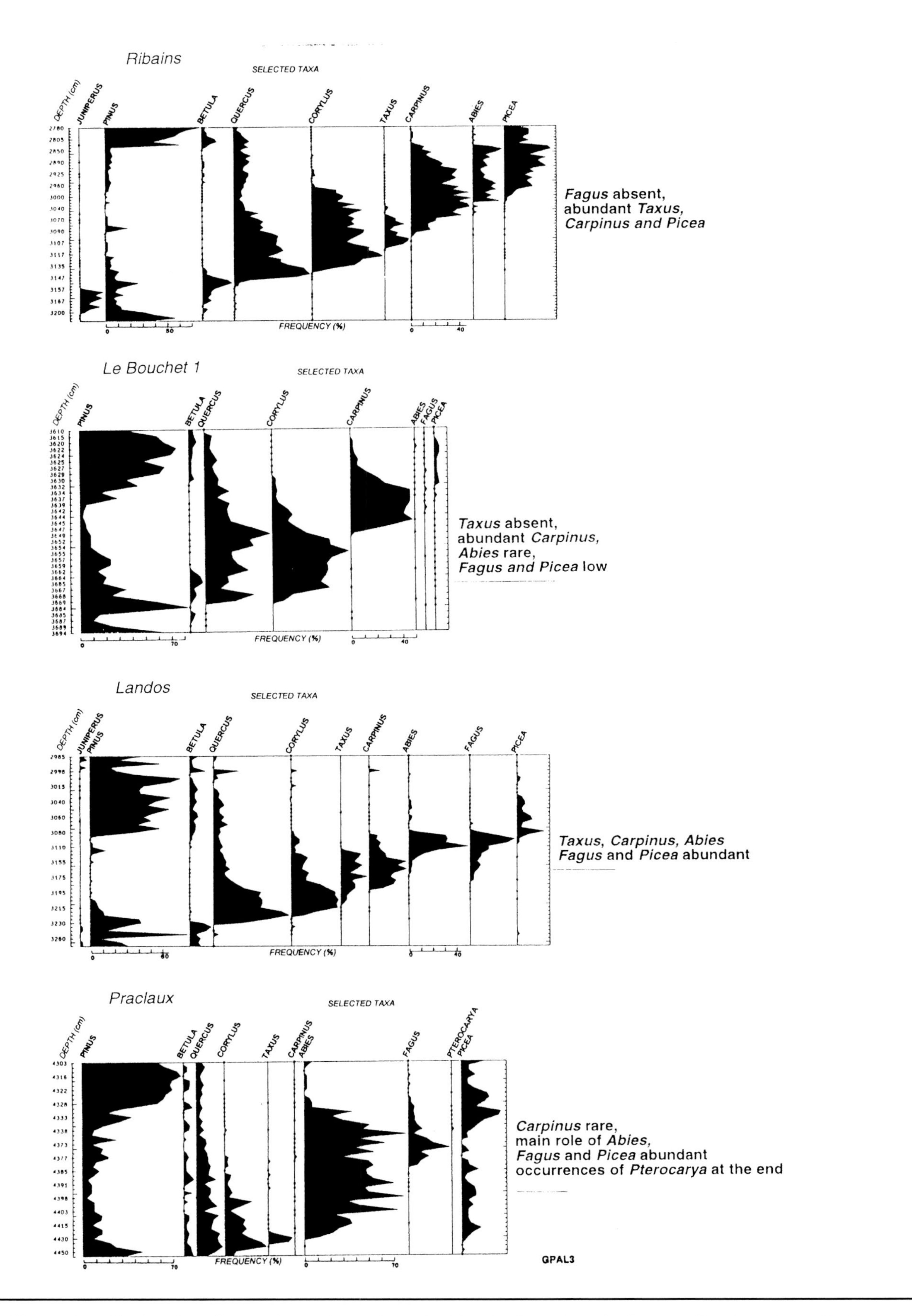

Fig. 3: Simplified pollen diagram for different interglacial episodes at Lac du Bouchet/Praclaux (magnified)

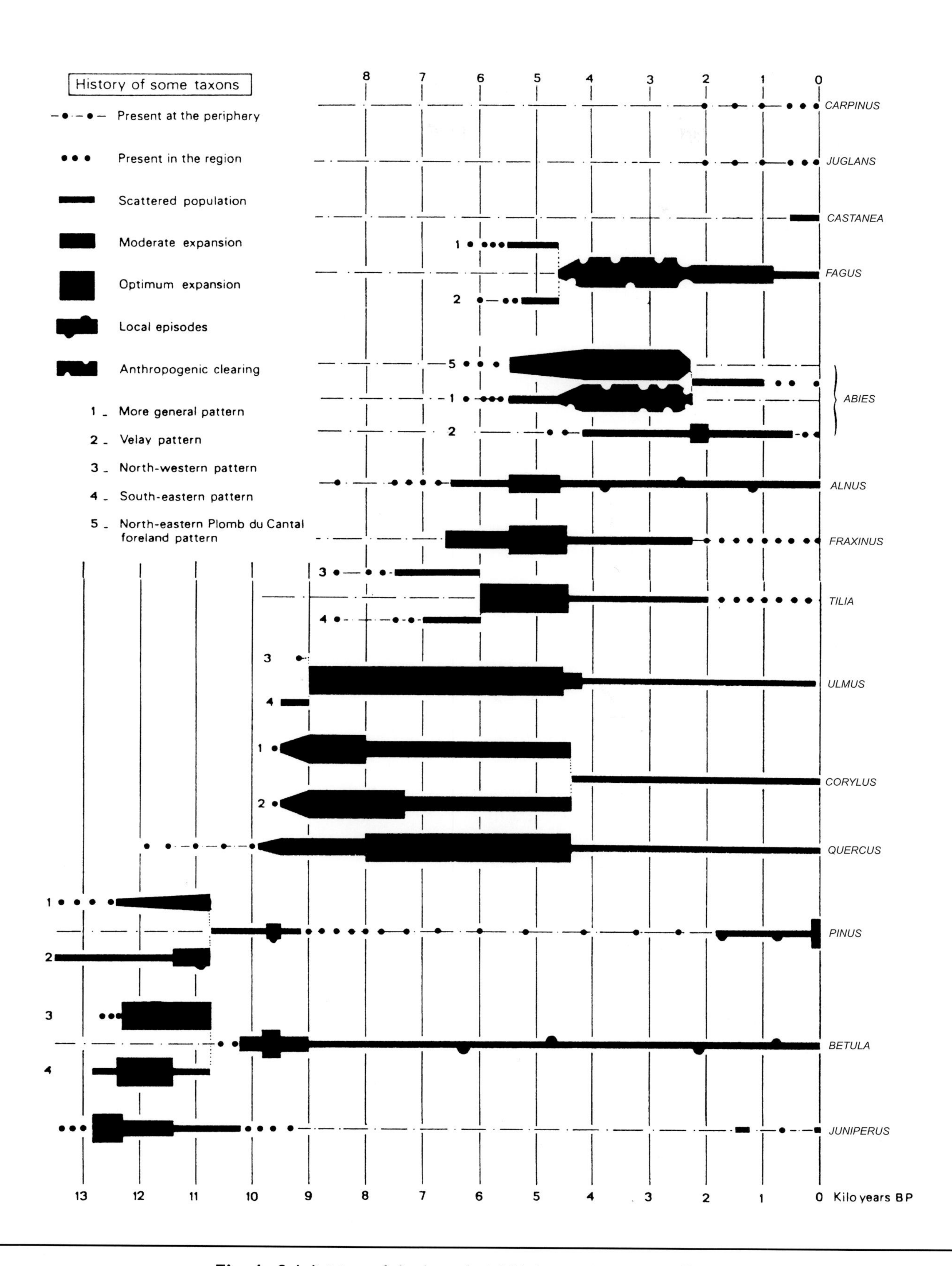

Fig. 4: Subdivision of the late-glacial Holocene in western Europe

Several generations of palynologists have analysed the history of vegetation in western Europe over the last fifteen millennia. They have proposed a general subdivision of the Holocene (Fig. 4) and have shown that forest types, just like human settlements, have developed along major migration paths, generally from the east to the west, but also from the south to the north.

These migration paths are well illustrated in the isopollinic maps of Huntley and Birks (1983). Since the publication of this book, many new sequences (dated by ^{14}C techniques) have been obtained which, along with advancing computer technology, led to the creation of the European Pollen Database (Cheddadi and de Beaulieu, 1995). The aim of this project was to gather and consolidate the numerous pollen data dispersed in different institutions so as both to conserve the data and to facilitate the best syntheses. Although this huge initiative is far from completion, the first paleovegetation maps showing paleobiomes at different time-slices in the past have already been produced (Prentice et al., 1996 and see also Berglund et al., 1995; Gliemeroth, 1995).

At the regional scale, we use small schema (called 'waste-steam pipes') to illustrate the main lines of the vegetation history (Fig. 5):

1. at the end of the last glaciation the number of pioneer heliophilous trees and shrubs increases;
2. during the Younger Dryas (ca. 10,500 BP) these modest woodlands retract;
3. during the Boreal and Atlantic periods (900 to 5,000 BP) there is an expansion of mixed oak forest (oak, elm, lime, hazel, and ash);
4. during the Sub-boreal period (5,000 to ca. 2,500 BP) the mixed oak forest is replaced by mountain forests of beech and fir;
5. in the Sub-Atlantic period (2,500 BP onwards) the forests retract due to human impact and pine woodlands develop as a result of secondary regrowth formation;

CALENDAR YEARS B.C	AGE B.P.	CHRONOZONES		CIVILISATIONS
		SUBATLANTIC	HOLOCENE	Gallo-Roman
				Iron Age
820	- 2700			
		SUBBOREAL		Bronze Age
3450	- 4700			
		Upper ATLANTIC		Neolithic
4880	- 6000			
		Lower ATLANTIC		
6900	- 8000			
		BOREAL		Mesolithic
8030	- 9000			
		PREBOREAL		
9200	- 10000			
		Younger DRYAS	LATEGLACIAL	Paleolithic
10970	- 11000			
		ALLERØD		
	- 12000			
		BØLLING		
13000	- 12700			
		Oldest DRYAS		
	- 14000			
16000	- 15000			
	- 16000	UPPER PLENIGLACIAL		
	- 17000			
19500	- 18000	GLACIAL MAXIMUM		

Fig. 5: History of the main forest taxa in the French Massif Central (after Magny, 1995)

6. in the late Sub-boreal and Sub-Atlantic period there is a large reduction of almost all the tree taxa.

These Holocene forest dynamics are representative of the middle mountain forest history in Central Europe. On the plains, mixed oak forest maintained itself until the Sub-Atlantic deforestation. In the schema presented here, the small depressions in the taxa curves have corresponded to temporary human action since about 4,500 BP.

■ THE MESOLITHIC

There is no evidence of human impact on the natural environment before the Mesolithic (Fig. 4). In France, Mesolithic populations do not appear to have had any impact on forests, but in the Alps archaeologists and palynologists have found evidence of temporary hunting settlements above the timberline that may have caused local impact on the forest (Fedele, 1992; Fedele and Wicks, 1996).

■ EARLY NEOLITHIC

The development of agriculture during the Neolithic produced both an increase in human population and deforestation. In the French Mediterranean region, at the time when the first farmers began to establish themselves, vegetation consisted mainly of deciduous oak forest (*Quercus pubescens* or Dusmast oak), stands of Mediterranean pine, and evergreen oak (*Quercus ilex*) in rocky habitats.

The first pollen records of the early Neolithic are dated to the second part of the seventh millennium (solar years: ca. 7500–7300 BP). Pollen analyses (Triat-Laval, 1978; Pons and Quézel, 1985; Reille and Pons, 1992, Reille et al., 1996) have shown that from the Neolithic onwards, forest burning gradually promoted the expansion of the sclerophillous *Quercus ilex*. This species was therefore wrongly identified by the first phytosociologists as the typical tree of the Sub-humid Mediterranean stage. In fact, *Quercus ilex* is usually a replacement tree, along with the Aleppo pine. Owing to repeated forest fires in many areas, the regressive vegetation of today consists mainly of the low bushes of the dwarf oak (*Quercus coccifera*) (Barbero et al., 1990). Bare soils are also visible.

Outside the Mediterranean region, the first early Neolithic forest clearings are only very rarely reflected in pollen diagrams that are derived from humid areas in which the regional forest development is recorded. In eastern France, the early phases of the Neolithic revealed by pollen records date to the beginning of the second part of the Older Atlantic period (ca. 5,500–5,400 CalBC).

In addition to the fall in arboreal pollen values and to the transitory increase of forest species indicative of open environments (i.e. birch and hazel), the presence of a few pollen grains belonging to species that indicate human activity is also evidence of deliberate forest clearing. The list of pollen taxa indicative of human activity is very similar throughout Europe (Behre, 1986).

Archaeological data suggest that increasing population density led to a clearing of the forest which brought about a disturbance in the dynamics of this ecosystem. However, quantitative analyses undertaken on the basis of multidisciplinary research in various parts of Europe (Frenzel, 1992) suggest that forest clearance did not exceed 5% of the total forest area.

■ MIDDLE NEOLITHIC–BRONZE AGE

Around 5,000 BP there is evidence of marked changes in European ecosystems. Many paleoecologists believe that it was not only climate change but also the clearance of the forest by Neolithic populations which created new conditions of competition that promoted the appearance of new forest types (i.e. the increase in beech coverage).

Pollen sequences derived from humid zones show that it is only after 4,500 BP (i.e. during the Middle and Upper Neolithic) that human impact on forests became evident throughout Europe—mainly from 'slash and burn' cultivation.

Deforestation was not uniform across Europe, but occurred in different stages. In many regions that were densely cultivated during the Roman Empire, a marked progression of the forest

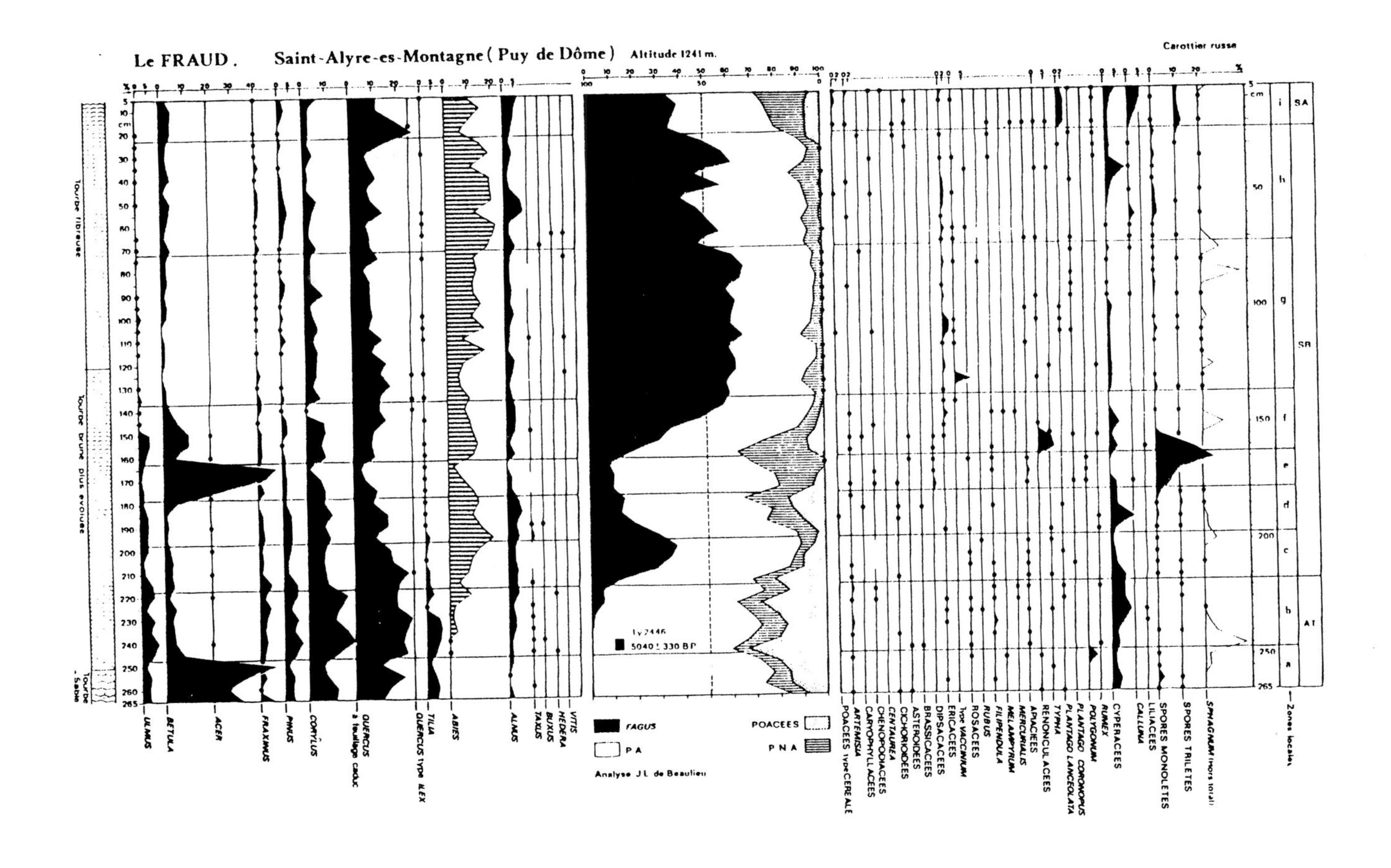

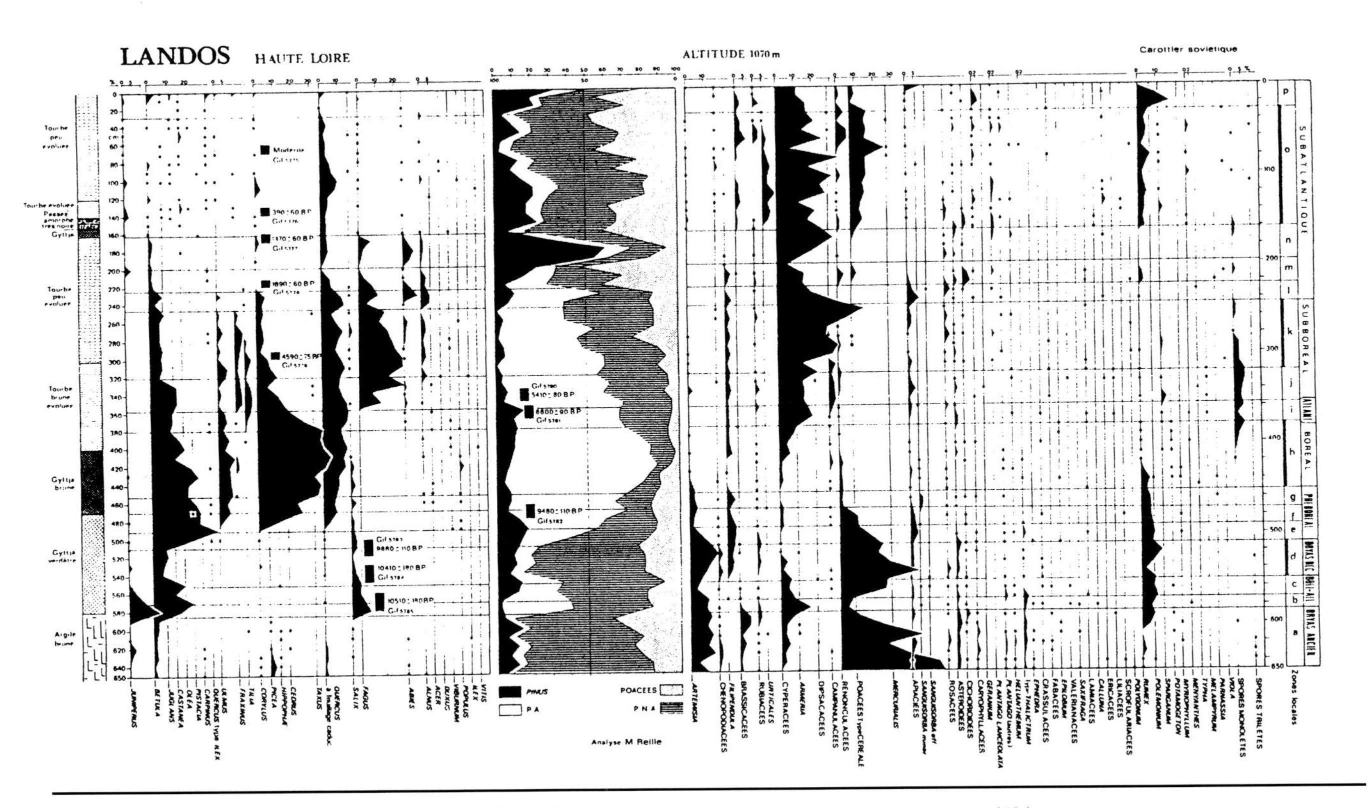

Fig. 6: Pollen diagrams from Le Fraud (6a) and Landos (6b)

occurred between AD fourth and ninth centuries, when the 'Barbarian' invaded from the east. However, deforestation has been most extensive in the Middle Ages (the climatic optimum of the tenth century being also marked by an increase in population) and in modern times. An awareness of the need to protect the forest wealth has only recently occurred.

This short synthesis can be illustrated by a few examples. In the French Massif Central, the Cezallier plateau is today totally deforested. The pollen diagram (Fig. 6a) obtained from Le Fraud peat-bog (Reille et al., 1985) shows a strong decrease in beech (which corresponds to major forest retraction) at about 4,000 BP followed by a period of new forest development. In fact, in this mountain zone, the extensive forest clearing that led to the present landscape does not seem to be older than the Medieval period.

In Velay, a pollen diagram (Fig. 6b) derived from the superficial peat of the Landos Crater (de Beaulieu et al., 1984) shows a marked decline in beech forest at the end of the Iron age and in Roman times. After the latter, the abandonment of cultivated land is accompanied by the development of replacement pine forests (indicated by high percentages of pine and the reappearance of beech). Deforestation then took place in the Middle Ages.

The Vallée des Merveilles in the Maritime Alps (above 2,000 m) contains many denuded rocks covered with engravings dated from the Bronze Age. Although the valley is probably not representative of intensive pastoral activity (de Lumley et al., 1976) palaeoecological studies have shown that almost everywhere in the Alps the timberline of the forested subalpine stage decreased in altitude by at least 200–400 m (Burga, 1988, David, 1993; Talon, 1997).

In Brittany (north-west France) which is famous for its great concentration of megalithic monuments, palaeoecological reconstructions from pollen analysis of peat-bogs and charcoal analysis from archaeological sites (Marguerie, 1992) highlight two major stages in the impact of human action on the environment in this area during the Holocene: forest clearance and the development of regressive heathland. In fact, exploitation of the forests and degradation of the soils are the prime causes of the development of the Brittany heathland. The average width of the growth rings in oak charcoal from domestic hearths shows a significant increase of between 5,200 and 2,000 BP (Fig. 7a). Similarly, more frequent use of oak firewood of small diameter from the branches of young trees can be observed. This implies that the environment became more and more open. Fig. 7b shows a strong growth in the charcoal of heathland taxa—heliophilic species such as broom, gorse, birch, and walnut—with a clear increase from the littoral to the inland areas. In the same way, many different taxa were used in domestic hearths during the Iron Age (Fig. 7c). This data suggests that the littoral areas were colonised first and that the increase in population induced farmers to move inland. Finally, it would appear that what we are observing is a tendency to economise on materials for construction purposes, and possibly the beginnings of forest management.

From the mid-Neolithic period to the end of the Bronze Age in the Alps and Jura foothills, the Palafittic civilisation settled on lake borders. At Lake Chalain, over fifty lake dwellings dating from these periods have been discovered. Each site shows a superimposition of several villages. A core taken at the centre of one of these sites contains evidence of three periods of human settlements stratified in the archaeologically sterile lacustrine chalk layers (Fig. 8). Dendrochronological and fine archaeological studies have enabled the dating of these settlement periods: level A/ was occupied between 2,700 and 2,630 CalBC, level A// was occupied between 2,800 and 2,700 CalBC, and level C (which comprises about 12 dwellings) was occupied between 3,030 and 2,920 CalBC.

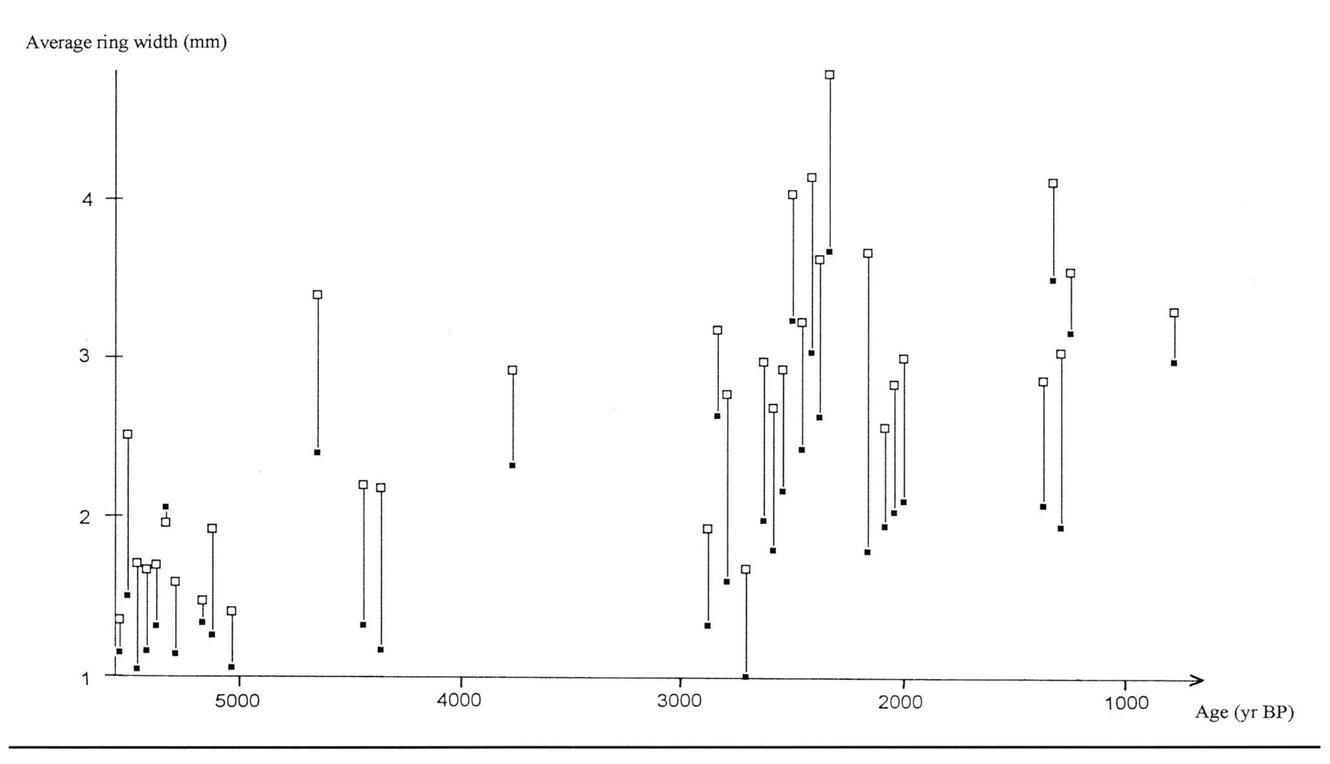

Fig. 7a. Evolution of the thickness of tree-rings on charcoals of large diameter oaks in domestic archaeological fires

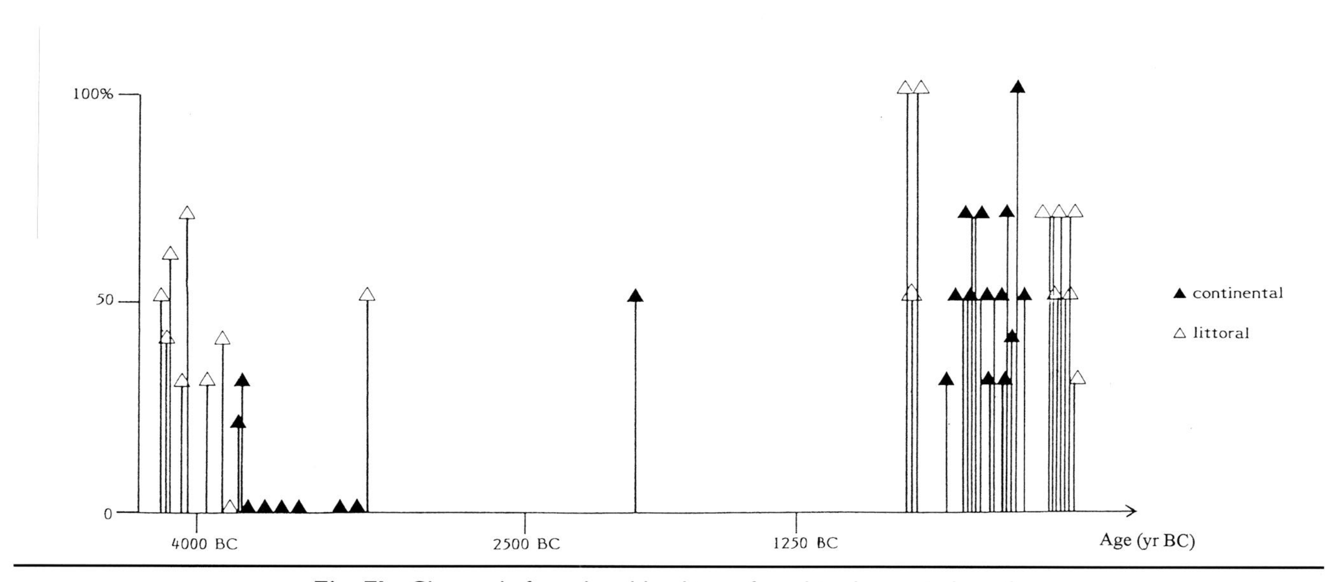

Fig. 7b. Charcoals from heathland taxa found in domestic hearths

A detailed pollen analysis of this core (Richard and Gery, 1993) reveals characteristic variations in the *Plantago lanceolata* and the *Plantago major/media* pollen curves (Fig. 8). At first, *Plantago lanceolata* is the dominant taxon but *Plantago major/media* gradually increases and becomes the most important taxon in the uppermost settlement phase.

Today in the Lake Chalain region, *Plantago lanceolata* largely grows on mowed meadows

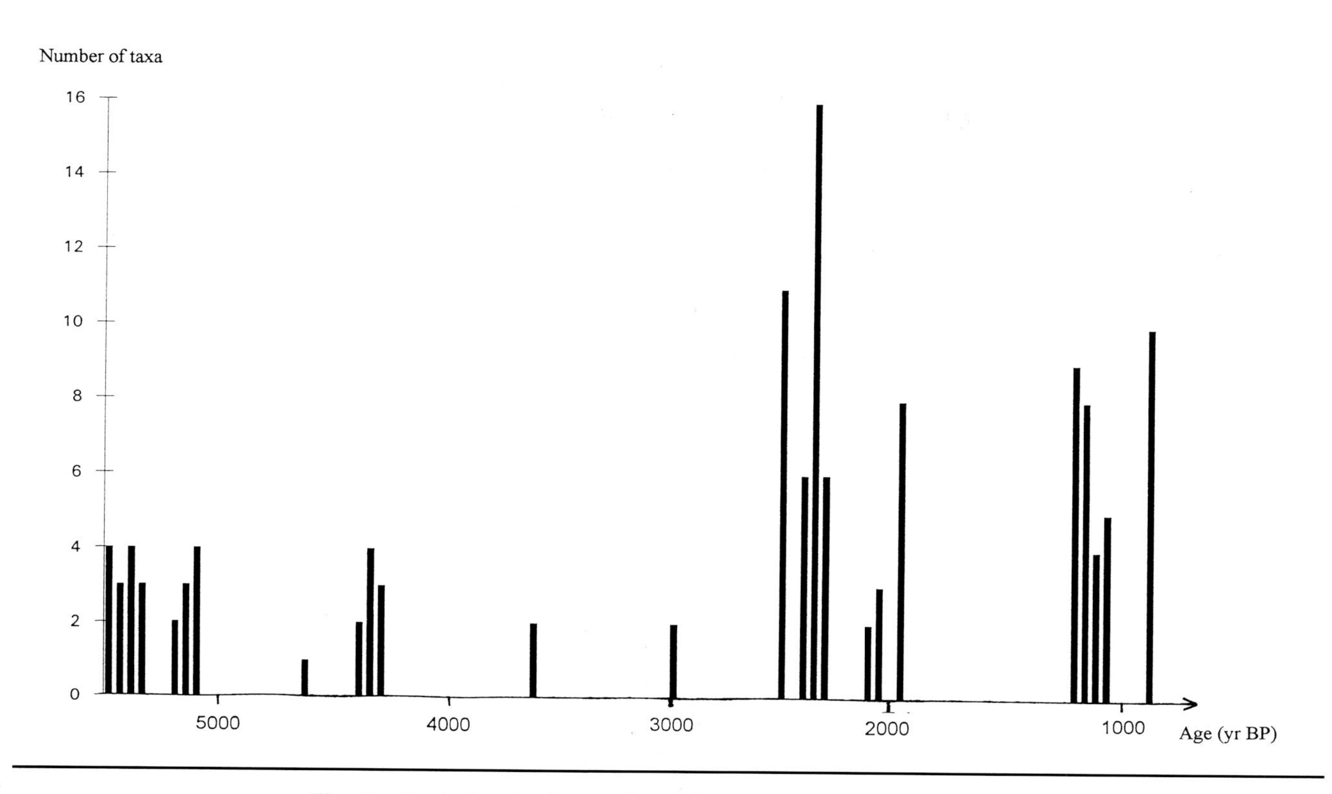

Fig. 7c: Evolution in the number of taxa used in domestic hearths

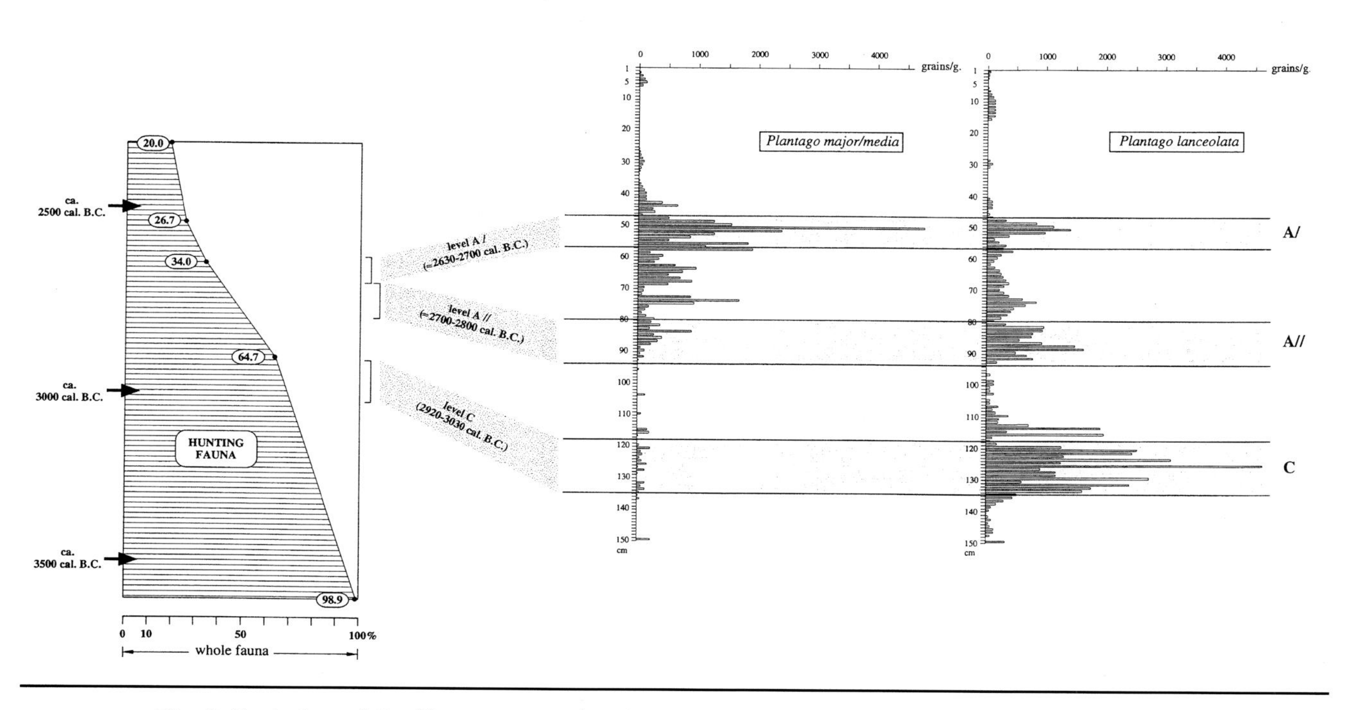

Fig. 8: Evolution of the Plantago major/media and Plantago lanceolata concentrations correlated with decreasing hunting fauna

whereas *Plantago major/media* is characteristic of trampled or over-trampled areas.

Archaeological studies (Pétrequin and Pétrequin, 1988) show that the demographic pressure must have increased between levels C and A// as the number of dwellings doubled and the land utilisation intensified accordingly. The archaeozoological studies of excavation sites at Lake Chalain and at nearby Lake Clarivaux (Pétrequin, 1989) reveal a substantial decrease in game animals and an increase in domesticated animals between 3,000 and 2,500 CalBC (Fig. 8). This suggests a change during the Neolithic from an economy based on hunting to one on animal breeding.

The shift in importance from *Plantago lanceolata* to *Plantago major/media* pollen representation is best explained as follows: between level C and level A//, substantial areas of meadows, pastures, or fallow lands came under a different management regime so that they were subject to trampling or over-trampling. This may have come about as a result of a considerable increase in the number of cattle and may also have led to an increase in deforested areas. At the same time, the abandoned cultivated areas were then without specific purpose in a primitive fallow system and were transformed into pastures. These changes favoured *Plantago major* and/or *media*, which is more resistant to trampling, to the detriment of *Plantago lanceolata*.

Using ecological, archaeozoological and archaeoecological arguments, it may be concluded that the rise in *Plantago major/media* and the decrease in *Plantago lanceolata* was due to an increase in the domestic grazing animal population. The fall of these two taxa may be explained by the development of the Neolithic agricultural system at about 2,800 CalBC, which involved a different equilibrium between arable farming, hunting, and stock rearing. On the other hand, dendroecological studies have also shown that during the same period the inhabitants of these villages had to travel several kilometres to find trees (especially oak) for construction material, which indicates a significant opening of the environment in the immediate surroundings of the lake.

The Perception of Forests by Man: The Myths

At the end of the Paleolithic (20,000 to 16,000 BP) during one of the coldest episodes that has ever existed on earth, there developed in southern France and in Spain a human civilisation that created the now famous rock art of this area.

Although humans were already dependent on wood for fire, only a few Norwegian pines and riverside trees, such as willows and birches, grew in the glacial steppe landscape. Nevertheless, it appears that vegetation did not play any role in the imagination of the Magdelanian people, or in their mythology, since only animals are represented in the art of this time.

However, with the advent of the Iron Age, it appears that Celtic and Gallic peoples belonged to a forest civilisation, even though they practiced farming. Druid priests celebrated their cult in oak forests. For example, in the Gallic sanctuary at the Seine springs, wooden statues of gods and votive offerings have been found. These so-called 'Barbarian' peoples were far less destructive than the Romans who constructed huge stone monuments and deforested large areas—even though Roman poetic and scientific works reveal a fondness for nature. In the Middle Ages, Genesis, which laid the foundations of Judeo–Christian thought, presents the earth as a garden offered to man so that he could mould it for his own use. At that time too, ancient legends resurged which presented the forest as a world of darkness inhabited by non-human creatures. Evangelisation by the great monastic orders was accompanied by extensive forest clearance. However, in some monasteries the forest was considered to be a renewable resource which should be carefully managed. To these monasteries we owe the survival of some of the most beautiful 'primitive' forests (de Beaulieu, 1977).

In France, the modern history of the forest is associated with Colbert, King Louis the fourteenth's minister, who created the National Office for Water and Forest in the seventeenth century. Colbert was not motivated by a concern for the environment, but rather the need for well-grown trees as material with which to build warships—owing to previous deforestation, suitable trees were rare at that time. A special body of engineers was created who were charged with collecting suitable tree specimens from around France or to acquire forests in which they grew.

After a certain decline during the French Revolution, the National Office for Water and Forest underwent considerable progress in the mid-nineteenth century, thanks to an engineer named Surrel. Surrel (1872) was convinced that the extent of deforestation in the French mountains, and principally in the southern Alps, meant that there was a severe risk of environmental catastrophe, especially flooding. His belief led to a long and impressive reforestation enterprise, as illustrated by comparisons between old photographs and the landscape of today (see Blanchard, 1956).

The lyricism with which Surrel promoted his cause shocked some scientists, led by Lenoble (1921). In Lenoble's opinion, the high altitude environments of the Mediterranean mountains were not all favourable for forestation. Subsequent palaeoecological studies have demonstrated not only that Lenoble was incorrect, but that deforestation started much earlier than Surrel believed.

The successful introduction of non-indigenous species by nineteenth and twentieth century forest engineers more concerned with productivity than maintaining the ecology, has altered natural biodiversity and upset the pre-existing order. The modern environmental movement has frequently reacted against such intervention, but some of their arguments are based, we believe, on such incorrect assumptions as:

1. the myth that forest exploitation necessarily means forest destruction. The only fir forest in the Durance Basin is the Boscodon (de Beaulieu, 1977), and it has been preserved by rational management and exploitation, first by monks and then by the state;
2. the myth of the 'primitive forest'. This term, generally used to justify reforestation, suggests a belief that no previous change occurred in forest cover prior to that which was produced by human impact. To the contrary, pollen analysis provides ample evidence of successive 'primitive' forests (sometimes over quite short time-intervals).

In south-western Europe today, the problem we have to face is not one of deforestation. Rather, it is the reduction in biodiversity that has resulted from a marked decline in traditional and extensive agriculture and the use of new heating methods. New dangers may occur, such as the increased fire risk in various regions (like the Provence) due to the expansion of secondary conifer forests that are very easily set alight. Despite increasing forest fires, in less than one hundred years the area covered by the Aleppo pine has increased from 35,000 to 200,000 ha and that of the Norwegian pine from 55,000 to 250,000 ha (Barbero and Quézel, 1990). Should we let these forests grow, or must we control their expansion? In view of these great changes, it may be that a promising prospect for the future is the young discipline of landscape ecology whose objective is to employ ecological, historical, and sociological approaches in order to develop a more harmonious management of our environment.

References

Barbero, M. and P. Quézel (1990): La déprise rurale et ses effets sur les superficies forestières dans la région Provence-Alpes-Côte d'Azur, *Bulletin Societé Linnéenne de Provence* 41: 77–88.

Barbero, M., G. Bonin, R. Loisel and P. Quézel (1990). Changes and Disturbances of Forest Ecosystems caused by Human Activities in the Western Part of the Mediterranean Basin, *Vegetatio*, 87: 151–73.

Behre, K. E. (1986): *Anthropogenic Indicators in Pollen Diagrams*, Balkema, Rotterdam.

Berger, A. (1977): Support for the Astronomical Theory of Climatic Change, *Nature*, 268: 44–5.

Berglund, B. E., H. J. B. Birks, M. Ralska-Jasiewiczowa, and H. E. Wright, (eds.), (1995): *Paleoecological Events During the Last 15 000 Years.*, Wiley & Sons, London.

Blanchard, R. (1956): *Les Alpes occidentales*, VII, Essai d'une synthèse, Arthaud edition, Grenoble.

Burga, C. (1988): Swiss Vegetation History During the Last 18,000 Years, *New Phytologist*, 110: 581–602.

Cheddadi, R. and J.-L. de Beaulieu (1995): European Pollen Database, *PAGES Workshop Report Series*, 9: 5–2, 57–60.

David, F. (1993): Evolution de la limite supérieure des arbres dans les Alpes françaises du Nord depuis la fin du glaciaire, Thèse, Univ. Marseille, 96pp.

de Beaulieu, J.-L. (1977). *Contribution pollenanalytique à l'histoire tardiglaciaire et holocène de la végétation des Alpes méridionales françaises*, Thèse ès Sciences, Université Aix-Marseille, 258pp.

de Beaulieu, J.-L., A. Pons and M. Reille (1984): Recherches pollenanalytiques sur l'histoire tardiglaciaire et holocène de la végétation des Monts du Velay (Massif Central, France), *Dissertationes Botanicae*, 72: 45–70 (Festschrift Max Welten).

de Beaulieu, J.-L., A. Pons and M. Reille (1988): Histoire de la flore et de la végétation du Massif Central (France) depuis la fin de la dernière glaciation. *Cahiers de Micropaléontologie*, N.S. 3 (4): 5–36.

de Lumley, H., M. E. Fontvielle, and J. Abelanet (1976). Les gravures rupestres de l'Age du Bronze dans la région du Mont Bégo (Tende, Alpes Maritimes), in E. Anati (ed.) *IX Congrés UISPP, Colloque XXVII, Les gravures protohistoriques dans les Alpes*, 7–35, ed. Louis-Jean, Gap, Fr.

Fedele, F. G. (1992). Steinzeitliche Jäger in der Zentral Alpen: Piano dei Cavalli (Splügen Pass). *Helvetia Archaeologica* (Zurich), 23 (89): 2–22.

Fedele, F. G. and L, Wicks (1996): Glacial/Postglacial Transition South of Splügen Pass (Northern Italy): Environment and Human Activity, Aiqua–PTSN Conference, Trento (Italy), Conference abstracts, 71–72.

Frenzel, B.(ed.) (1992): Evaluation of Land Surfaces Cleared from Forests by Prehistoric Man in Early Neolithic Times and the Time of Migration Tribes, *Paläoklimaforschung*, 8: 225.

Gliemeroth, A. C. (1995): Paläoökologische Untersuchungen über die letzten 22,000 Jahre in Europa, *Paläoklimaforschung*, 18: 252.

Huntley, B. and H. J. B. Birks (1983): *An Atlas of Past and Present Pollen Maps for Europe*: 0–13,000 years ago, Cambridge University Press, Cambridge.

Lenoble, F. (1921): Les limites de végétation de quelques espèces méditerranéennes dans le bassin moyen du Rhône et les Préalpes sud occidentales, *Revue de Géographie Alpine*, 9: 457–70.

Magny, M. (1995): *Une histoire du climat. des derniers mammouths au siècle de l'automobile*, Errance, Paris.

Marguerie, D. (1992): Evolution de la végétation sous l'impact humain en Armorique du Néolithique aux périodes historiques, *Travaux du Laboratoire d'Anthropologie de Rennes*, 40: 313.

Petrequin, P. (1989): *Les sites littoraux néolithiques de Clairvaux-les-Lacs (Jura), Tome II. Le Néolithique Moyen*, Maison des Sciences de l'Homme, Paris.

Petrequin, A. M. and P. Petrequin (1988): *Le Néolithique des lacs, Préhistoire des lacs de Chalain et de Clairvaux (4000–2000 av. J. C.)*, Errance, Paris.

Pons, A. and P. Quézel (1985): The History of the Flora and Vegetation and Past and Present Human Disturbance in the Mediterranean Region, in C. Gomez-Campo (ed.), *Plant conservation in the Mediterranean area*, W. Junk, Dordrecht, pp. 24–43.

Prentice, J. C., J. Guiot, B. Huntley, D. Jolly, and R. Cheddadi (1996): Reconstructing Biomes From Paleoecological Data: A General Method and Its Application to European Pollen Data at 0 and 6 Ka, *Climate Dynamics*, 12: 185–194.

Reille, M., V. Andrieu and J.-L. de Beaulieu (1996): Les grands traits de l'histoire de la végétation des montagnes méditerranéennes occidentales, *Ecologie*, 27 (3); 153–69.

Reille, M., J.-L. de Beaulieu, and A. Pons (1985): Recherches pollenanalytiques sur l'histoire tardiglaciaire et holocène de la végétation du Cézallier, de la planèze de St-Flour et de la Margeride (Massif Central, France), *Pollen et Spores* 27(2): 209–70.

Reille, M. and J.-L. de Beaulieu (1995): Long Pleistocene Pollen Records from the Praclaux Crater, South-Central France, *Quaternary Research*, 44: 205–15.

Reille, M. and A. Pons (1992): The Ecological Significance of Sclerophyllous Oak Forests in the Western Part of the Mediterranean Basin: A Note on Pollen Analytical Data, *Vegetatio*, 99–100: 13–17.

Richard, H. and S. Gery (1993):Variations in Pollen Proportions of *Plantago lanceolata* and *Plantago major/media* in a Neolithic Lake Dwelling, Lake Chalain, France, *Vegetation History and Archaeobotany*, 2: 79–88.

Surrel, A. (1872): *Etude sur les torrents des Alpes*, Dunod edn., Paris.

Talon, B., (1997): *Evolution des zones supra-forestières des Alpes sud-occidentales françaises au cours de l'Holocène, Analyse pédoanthracologique*, Thèse en Sciences, Université Aix-Marseille, III, 207pp.

Triat-Laval, H. (1978): *Contribution pollenanalytique à l'histoire tardi- et postglaciaire de la végétation de la basse vallée du Rhône*, Thèse ès Sciences, Université Aix-Marseille.

CHAPTER 3

Forest and Civilisation on Easter Island

J. R. FLENLEY

There have been three great controversies about the vegetation of Easter Island. First, was the island ever forested? Second, if it was forested, why did the forests disappear? Third, could the disappearance of the forests be responsible, even partly, for the collapse of the Easter Island civilisation?

Let us take these matters in turn. There is no historical evidence that the island was ever forested. In the eighteenth century, the earliest visitors described seeing 'woods' in the distance, but these were probably the low scrub formed by the *toromiro* (*Sophora toromiro*) and were perhaps only 2 m in height (Cook, 1777). Furthermore, the present flora of the island (Skottsberg 1956; Zizka, 1991) contains only the *toromiro* and two shrubs (*Caesalpinia bonduc* and *Lycium carolinianum*). This is far less than the rich tree diversity of most tropical or subtropical islands, a peculiarity explained by van Balgooy (1960) as the result of Easter Island's extreme isolation.

However, there is considerable indirect evidence that Easter Island was forested in the past. In most parts of the world, the major determinant of vegetation is climate. The present climate of Easter Island is admirably suited to forest growth: it is warm and moist, with only mild seasonality. It is true that occasional droughts have been recorded, but planted trees over a hundred years old survive on the island. In most parts the soils are well suited to tree growth, especially on Poike where the greater geological age (ca. 3 million years) has given time for deep weathering of the volcanic rocks to produce a thick fertile soil.

There is now strong evidence that the native tree flora was previously more extensive and varied. Years ago, Edmundo Edwards, a resident of the island, found large quantities of nuts in a lava cave. He collected a large number and submitted some for identification. They were identified as *Calophyllum inophyllum*, a tree that occurs on the tropical Pacific Islands. Similar nuts had been recorded earlier by Skottsberg, being worn as ornaments and used by children as tops. Skottsberg had identified these as *Thespesia populnea*, a tree that grows on the island but is believed to have probably been introduced (Skottsberg, 1956). However, both these identifications were probably wrong. Eventually, further nuts were found in lava caves by the French speleologists J. Groult and A. Gautier (Gautier and Carlier, 1987) and were submitted to Dr John Dransfield at Kew. He identified them correctly as palm nuts of an extinct endemic species *Paschalococos disperta*, closely related to the Chilean wine palm, *Jubaea chilensis* (Dransfield et al., 1984; Dransfield in Zizka, 1991). The Chilean wine palm is the largest palm in the world (Grau, 1994) and is a food source. The nuts are edible and the sweet sap can be used to make palm sugar and palm wine. It seems likely that the Easter Island palm was both large and useful. In 1868, Palmer saw stumps of the last palm trees (Palmer 1870a, b), so the extinction could have been very recent. Casts of these trees—trunks and petioles—have been found in volcanic lava, just above a buried soil, in which they were presumably growing. The trunks appear to have been about 0.5 m in diameter and at least 6 m long.

Palm nuts have now been found in caves at three different parts of the island (J. Vignes, pers. comm.) and also in the exposed alluvium at valley mouths (Orliac, 1993). One was found inside an *ahu* (statue platform) (Flenley et al., 1991) and fragments of palm nut shell were abundant in the archaeological excavation at Anakena (S. Rapu, pers. comm.) indicating that it was both abundant and widespread. This is borne out by the discovery of carbonised root channels on the slopes of Terevaka and well exposed by soil erosion on Poike. The diameter of the channels and their lack of branching is consistent with a palm of the *Jubaea* type. Some of the palm nuts have been dated by radiocarbon assay to 820 +/- 40 years BP, indicating that the palm was alive and well until after people arrived on the island around 1,600 years ago.

Sandalwood (*Santalum* sp.) may also have existed on Easter Island at one time. The evidence for this is the beautiful carved wooden hand collected on the island by Captain Cook and preserved in the British Museum. The wood has now lost its distinctive perfume, but this may be because of its great age. Unless it is incorrectly attributed to Easter Island, or incorrectly identified botanically, it is good evidence of the former occurrence of this tree on the island.

In recent years, palynology has greatly augmented our knowledge of Easter Island's extinct vegetation. Fossilised palm pollen (presumably *Paschalococos disperta*) was reported by Selling in samples from Rano Kau collected by the Heyerdahl expedition (Heyerdahl and Ferdon, 1961). Later work by Flenley et al. (1991) has shown that palm pollen was frequent in all three craters, and was usually the dominant type at the two lower altitude sites (Rano Kau and Rano Raraku). From the pollen record, it is inferred that at Rano Raraku the vegetation was continuously dominated by trees, mostly palms, for over 30,000 years (Fig. 1). Only in the last 1,200 years did the palm decline and become extinct. This is significant because there have been major global climate changes in the last 30,000 years, including climates both warmer and colder than today. Clearly the palm was not a species likely to be eliminated by a minor climatic fluctuation and therefore its recent disappearance cannot be attributed to climatic change alone.

The pollen record shows that there were other species too. *Coprosma*, a shrub or small tree found on many Pacific islands, also used to grow on Easter Island. So did a species of the Myrtaceae, possibly the *rata* tree, *Metrosideros* sp. The *hau* tree, *Triumfetta semitriloba*, used for making rope and considered by Skottsberg (1956) to have been introduced, has been shown to be a native, having occurred on the island for 30,000 years. There was also a species of daisy tree (*Asteraceae*) similar to those found in the Marquesas.

Some people have wondered whether pollen grains could have been carried on the wind from other islands or from South America. Fortunately, this idea can be tested by examining the deposits on Salas-y-Gomez, an island relatively near to Easter Island (Flenley et al., 1996). The pollen grains found on Salas-y-Gomez are derived almost exclusively from its few indigenous species, so transoceanic transport was therefore unlikely.

We may therefore conclude that Easter Island was formerly forested, and that this forest continued with little change from 30,000 years ago until about 1,200 years ago (see Fig. 2). The dominant tree was probably the Easter Island palm at lower altitudes, and the *Coprosma* and daisy tree above about 400 m. By about 600 years ago (say AD 1400) the forest, at least at Rano Kau,

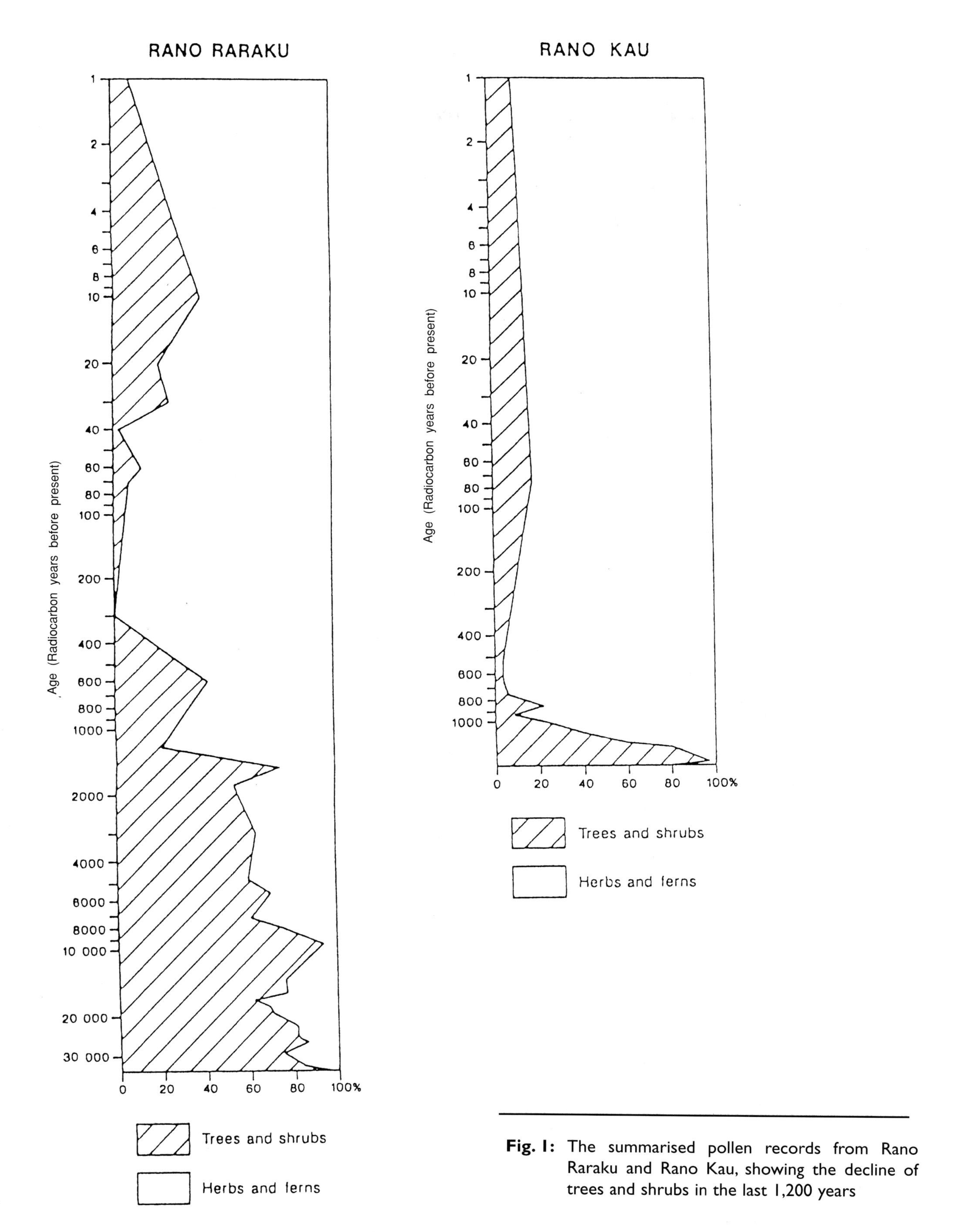

Fig. 1: The summarised pollen records from Rano Raraku and Rano Kau, showing the decline of trees and shrubs in the last 1,200 years

Crater lake Rano Kau. Photo by Yasuda

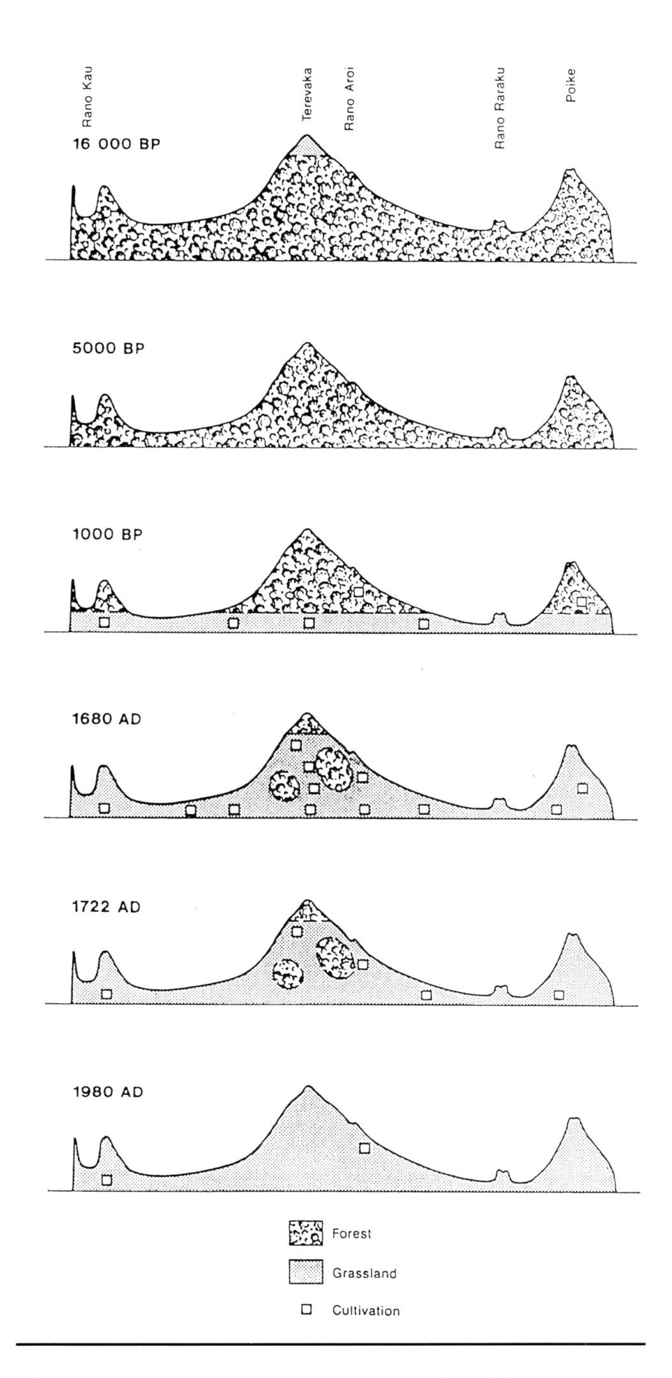

Fig. 2: The vegetation of Easter Island at different times, according to the pollen evidence from Rano Raraku, Rano Kau, and Rano Aroi

was gone. At least three possible causes of forest decline have been suggested: climate change, volcanic eruption, and human impact. In favour of climate change, it must be said that the island is today subject to occasional severe droughts. Perhaps during the sixteenth to nineteenth centuries, when the rest of the world experienced the 'Little Ice Age', droughts on the island may have been more severe (McCall, 1993). Against this it must be pointed out that the island remained forested throughout the major Ice Age which occurred 18,000 years ago (see Fig. 1). It therefore seems unlikely that the 'Little Ice Age' could have destroyed Easter Island's forests.

The island is volcanic but had there been major volcanic eruptions in the last few thousand years, there should have been volcanic ash in the Rano Raraku deposits. Instead, only two very thin layers of ash were found and these dated to ca. 12,000 BP, long before deforestation.

An important point in favour of the 'human impact' theory is that all the palm nuts so far found in caves have been gnawed by rats. The Polynesian rat, *Rattus concolor*, was introduced everywhere by Polynesian voyagers as a source of food. It frequently had a disastrous effect on plants and on seabird colonies. There is evidence from bird bones in the Anakena deposits of a great decline in sea bird numbers and diversity on Easter Island (Steadman, 1993). The rats may have had a similar effect, as they ate palm nuts and thus prevented regeneration.

The human population would have destroyed the forest by felling and burning both to clear land for crop cultivation and to obtain firewood. The palynological record shows that forest decline was accompanied by the presence of charcoal. Although natural fires are a possibility, it would be a curious coincidence if their increase in frequency was not related to the arrival of people. Overall, the balance of evidence favours human action as the cause of forest decline. The decrease in manure from the declining numbers of sea birds may also have affected forest growth.

This brings us to the third controversy: could forest decline have been responsible for the collapse of the Easter Island civilisation? We know from legends that the civilisation collapsed in war and famine around AD 1680 (Heyerdahl and Ferdon, 1961). We know that the lowland

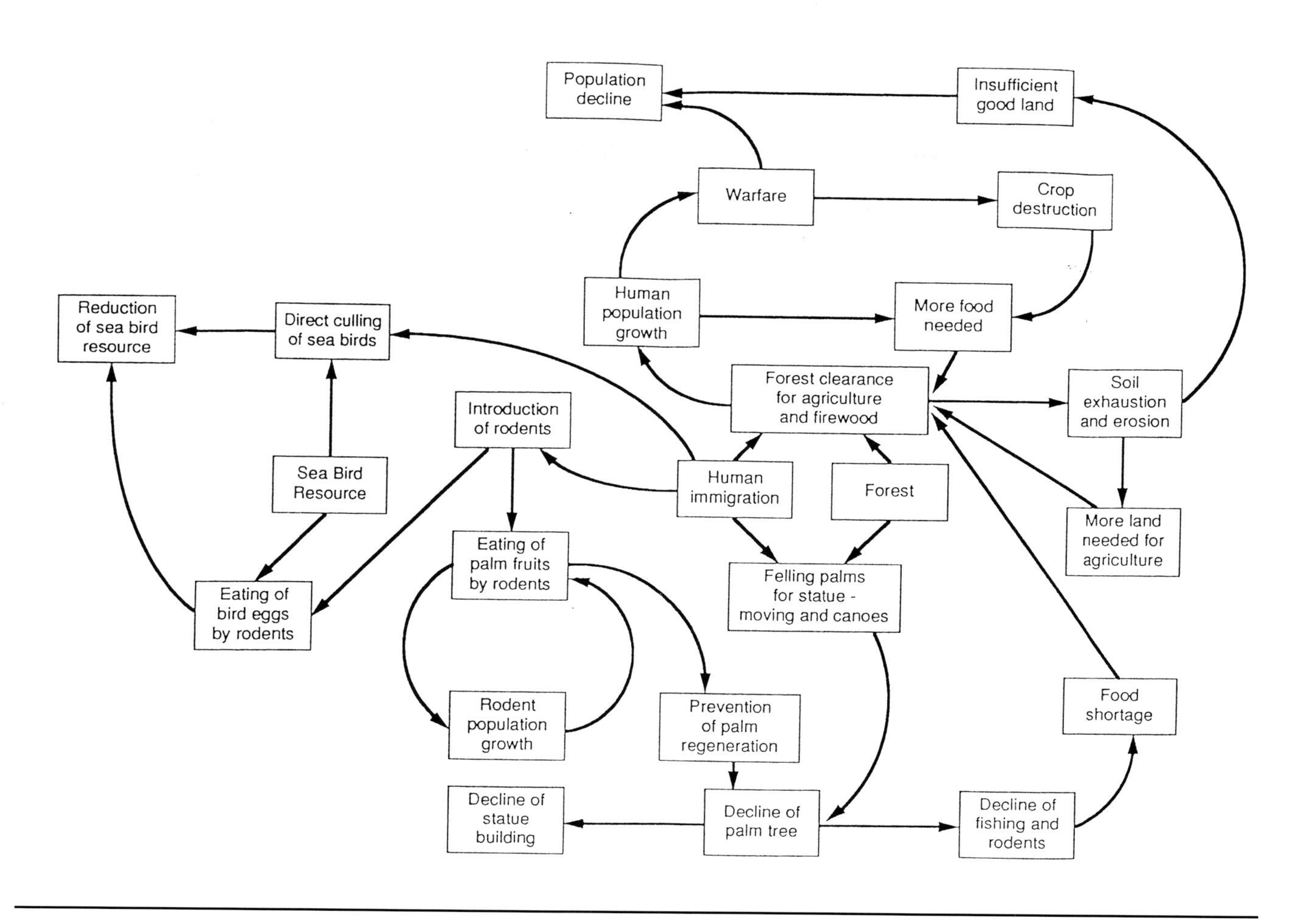

Fig. 3: A hypothetical model showing how an ecological disaster could have occurred on Easter Island

forests were exhausted by about AD 1400. It is possible that the upland forests could have lasted a further two centuries, but the major deforestation took place *before* the collapse of the civilisation. This is logical: a cause must precede an effect. Furthermore, it is possible to suggest a sequence of events that would have led to ecological collapse on a small island with a burgeoning population. The ever increasing need for food, firewood, and materials for canoe-building and statue-moving would have exerted immense pressure on the remaining forests. Soil erosion probably set in on overused land. In contrast with the more equatorial islands, there was no coral reef to trap the eroded soil, which was therefore lost. The fishing resource was reduced as canoes inevitably became smaller. Rats had all but eliminated the seabird resource. Without palm tree rollers, statues became immovable. The whole concept is of a series of interactions, summarised in Fig. 3, which inevitably led to population decline and the collapse of the Easter Island civilisation.

There are several other theories about the mysterious disaster of Easter Island, but most have little supporting evidence (Bahn and Flenley, 1992). We shall probably never know what exactly happened. However, it is certainly possible, as McCall (1993) has suggested, that climatic change may have been involved. If there were droughts in the seventeenth century, which is quite likely, they may have given the *coup de grâce* to an agricultural system already stretched to breaking point for ecological reasons.

References

Bahn, P. and J. R. Flenley (1992): *Easter Island, Earth Island*, Thames & Hudson, London, 240pp.

Balgooy, M. M. J. (1960): Preliminary Plant Geographical Analysis of the Pacific as Based on the Distribution of Phanerogam Genera, *Blumea*, 10: 385–430.

Cook, J. (1777): *A Voyage Towards the South Pole, and Round the World, 1772–75*. 2 vols., Strahan & Cadell, London, 368 and 378pp.

Dransfield, J., J. R. Flenley, S. M. King, D. D. Harkness and S. Rapu (1984): A Recently Extinct Palm From Easter Island. *Nature*, London, 312: 750–52.

Flenley, J. R., L. K. Empson and G. Velasco (1996): Salas-y-Gomez: A Natural Pollen Trap in the Pacific and its Significance for the Interpretation of Island Pollen Diagrams, *Rapa Nui Journal*, 10: 17–20.

Flenley, J. R., A. S. M. King, J. T. Teller, M. E. Prentice, J. Jackson and C. Chew (1991): The Late Quaternary Vegetational and Climatic History of Easter Island, *Journal of Quaternary Science*, 6: 85–115.

Gautier, A. and P. Carlier (1987): Les caverns de l'Ile de Paques, *Archaeologia*, 221: 34–47.

Grau, J. (1994): *Jubaea, the Palm of Chile and Easter Island?* Proceedings International Congress of Ecology, Manchester, UK, Aug. 1994.

Heyerdal, T. and E. Ferdon Jr. (eds) (1961): *Reports of the Norwegian Archaeological Expedition to Easter Island and the East Pacific*, 2 vols, Allen & Unwin, London.

McCall, G. (1993): Little Ice Age: Some Speculations for Rapanui, *Rapa Nui Journal*, 7: 65–70.

Orliac, M. (1993): Le palmier des Pascuans, pp. 403–13 in *Les mysteres résolus de l'Ile de Paques*, Editions Step, Paris, 510pp.

Palmer, J. L. (1870a): A Visit to Easter Island, or Rapa Nui, *Proceedings of Royal Geographical Society*, 14: 108–119.

Palmer, J. L. (1870b): A visit to Easter Island, or Rapa Nui, in 1868. *Journal of Royal Geographical Society*, 40: 167–81.

Skottsberg, C. (1956): *The Natural History of Juan Fernandez and Easter Island*, Vol. 1, Almquist & Wiksells, Uppsala, 438pp.

Steadman, D.W. (1993): Extinct Birds from Easter Island, Abstracts International Conference on Easter Island Research, Laramie, Aug. 1993, p. 35.

Zizka, G. (1991): Flora of Easter Island (flowering plants), *Palmarum Hortus Francofurtensis*, 3: 1–108.

CHAPTER 4

Easter Island: Its Rise and Fall

PAUL BAHN

Easter Island is a tiny volcanic speck in the South Pacific measuring 64 square miles. Located at a distance of 2,340 miles from South America and 1,400 miles from the nearest island, Pitcairn, to the north-west, this is the most isolated area of permanently inhabited land. It is famous for its astonishing Stone Age culture which produced hundreds of enormous stone statues, many of them set up on massive stone platforms. Today, the whole of Easter Island constitutes a gigantic open-air museum: Rano Raraku quarry and Rano Kau crater are two of the most spectacular sights in world archaeology.

According to radiocarbon dates, Easter Island's structures were already well developed by AD seventh century and most researchers believe that the colonists arrived in the early centuries AD (Bahn and Flenley, 1992). The island is so remote and isolated that it is unlikely that there was ever more than one major influx of people by canoe, and the island's archaeological record certainly suggests a single unbroken development of material culture. Once settled on the island, the colonists were trapped there, and it constituted their whole world.

The earliest known contact with the outside world occurred on Easter Sunday 1722, when the Dutch navigator Jacob Roggeveen encountered and christened the island, and described its inhabitants. Subsequent eighteenth century visitors included such famous explorers as Captain Cook from Britain, in 1774, and the Comte de La Pérouse from France, in 1786. Archaeological investigation began sporadically in the late nineteenth century, but entered a new phase with the Norwegian expedition of 1955, led by the adventurer Thor Heyerdahl. He brought archaeologists with him, most notably the American William Mulloy, who laid the foundations for the research and restoration that continue today. This expedition carried out the first stratigraphic excavations, and obtained the first radiocarbon dates and pollen samples, besides conducting valuable experiments in carving, moving, and erecting statues.

For decades, Thor Heyerdahl has been promulgating the theory that Polynesia in general, and Easter Island in particular, was first settled by South American Indians, and only subsequently by Polynesians. In an attempt to prove that this was possible, in 1947 he caught the world's imagination by floating on a balsa-wood raft, the famous *Kon-Tiki*, from Peru to the Tuamotu Islands, east of Tahiti. Unfortunately, this exploit proved nothing more than that it is possible to survive such a voyage. Heyerdahl's theory has found no support whatsoever in archaeological research. The island's material culture shows absolutely no links with the New World beyond a few superficial similarities, and indeed displays a total absence of the most characteristic features of South American prehistoric cultures (pottery, pressure-flaking of stone tools, weaving, etc.). In contrast, every aspect of Easter Island's archaeology and anthropology—architecture, technology, religion, settlement patterns, art, anatomy, blood groups, DNA, and linguistics—points unequivocally to the north-west (Polynesia, and most probably Mangareva) as the source of its population and culture.

In its 3 million years of existence, Easter Island—or Rapa Nui (big Rapa) as it is now called by its inhabitants—had slowly developed a balanced ecosystem. The original volcanic rocks have weathered to provide a fertile soil, enriched by the guano of thousands of sea- and land-birds. Plants have immigrated—some brought by birds, some by the wind and some floating on the sea. As we now know from analyses of pollen and the remains of a now extinct snail, the island once supported a diverse rain forest dominated by a great palm tree similar to *Jubaea chilensis*, the Chilean wine palm, the largest in the world.

This idyll was disturbed by the arrival of the Polynesian voyagers: probably a few dozen men, women and children in one or more large double canoes, bringing with them the domestic animals (chickens, rats, pigs, and dogs) and food plants (bananas, sweet potatoes, taro, breadfruit) with which the Polynesians transformed the environment of so many Pacific islands. Pigs and dogs, if they ever arrived, did not survive long on Easter Island, and breadfruit could not grow there. Although few in number at first, the colonists set about clearing the forest to plant their crops. The native birds, unused to humans, fell an easy prey to hunters (Steadman, 1995), and the rats stole their eggs, so the few remaining seabirds retreated to offshore islets. Indeed, we only know of their former existence on the main island from the bones left in lava caves or archaeological sites. During this early phase, the islanders seem to have constructed simple types of *ahu* (platforms), with small and relatively crude statues on or in front of them.

The second phase of Easter Island's history, from ca. AD 1000 to 1500, was its 'golden age' when tremendous energy was devoted to the construction of more and bigger ceremonial platforms (rubble cores encased in well-cut slabs) and hundreds of large statues. As the human population throve in this earthly paradise, numbers gradually increased, perhaps reaching a peak of 10,000 or even 20,000 around AD 1500. This put pressure on the supply of land, and the inevitable decline of the forest can be seen clearly in the record of fossilised pollen from the island's crater swamps (Flenley, 1993; Flenley et al., 1991).

At least 800 *moai* (statues) were carved, almost all of them in the soft volcanic tuff of the Rano Raraku crater, by means of basalt hammerstones. All were variations on a theme: a human figure with prominent angular nose and chin, and often elongated perforated ears containing discs. The bodies, which end at the abdomen, have arms held tightly to the sides, and hands held in front, with elongated fingertips meeting at a stylised loincloth. Over 230 statues were transported considerable distances from the great Rano Raraku quarry to platforms around the edge of the island, where they were erected, their backs to the sea, watching over the villages clustered around each platform. They are believed to represent ancestor figures. It was traditionally thought that statues were dragged horizontally to their destinations, but recent experiments suggest that the most efficient mode of transportation was upright, on a sledge and rollers.

At the most important and prestigious platforms, statues were given eyes of white coral and a separate *pukao* (topknot) of red scoria was raised and placed on the head (Pavel, 1995). The statues placed on platforms vary from 6 to 33 feet in height, and weigh up to 82 tons. There might be up to 15 in a row on a single platform. The quarry however still contains at least 394 statues at every stage of manufacture; one of them, 'El Gigante', is 65 feet long and would have weighed up to 270 tons.

The third and final phase of the island's prehistory saw the collapse of the earlier way of life: statues ceased to be carved, cremation gave way to burial. What seem to be 1,000 years of peaceful coexistence were shattered by the manufacture in huge quantities of *mataa*, spearheads, and daggers of obsidian, a sharp black volcanic glass. Conflict led to the toppling of the statues, and was resolved by an apparent abandonment of a religion and social system based on ancestor worship in favour of one featuring a warrior élite.

An annual chief or 'birdman' was chosen each year at the ceremonial village of Orongo, its drystone corbelled houses perched high on the cliff between the Rano Kau crater and the ocean. Each of the candidates—mostly ambitious warlords from dominant or victorious clans rather than hereditary aristocrats—had a young man to represent him. Each September (springtime in this part of the world) these unfortunate young men had to make their way down the sheer 1,000-foot cliff to the shore, and then swim over a mile on a bunch of reeds through shark-infested swells and strong currents to the largest and outermost islet, Motu Nui. Here they awaited—sometimes for weeks—the arrival of a migratory seabird, the Sooty Tern. The aim was to find its first egg. The winner would shout the news to his employer on the Orongo clifftop, and then swim back with the egg securely held in a headband. His master now became the new sacred birdman, shaved his head, eyebrows, and eyelashes, and had his head painted. He went off to live in lazy seclusion for a year, and refrained from washing, bathing, or cutting his nails. Orongo's rich rock art is festooned with carvings of the birdmen, sometimes holding the Sooty Tern egg which symbolised fertility. This was the system that was still developing when the Europeans arrived, and which ended with the arrival of missionaries in the 1860s.

The causes of the island's decline and change were probably complex, but can ultimately be traced to one major factor: pollen analysis has shown that Easter Island has the most dramatic record of deforestation in the archaeological record (Flenley, et al., 1991; Bahn and Flenley, 1992; Flenley, 1993; Flenley, this volume). From at least 1,200 years ago, one can see a massive reduction in forest cover, until, by the time of the arrival of Europeans, there were virtually no large trees left. The imported Polynesian rats fed on the palm fruits and helped prevent regeneration. Without the palm tree and other timbers, statues could no longer be moved; ocean-going canoes could no longer be built, thus cutting the population off from the crucial protein-supply of deep-sea fish (the island has no coral reef or lagoon). Deforestation also caused massive soil erosion which damaged the island's crop-growing potential. Chickens became the most precious source of protein.

It is impossible to know exactly what happened on Easter Island, but the probably steady growth of population—perhaps up to 20,000 or even more—together with the decline in food and the increasing importance of essentially useless activities (platform building, statue carving, and transportation) clearly led to a collapse. Starvation led to the rise of raiding and violence, perhaps even to cannibalism, though the only evidence for this lies in oral traditions (Bahn, 1997).

By 1722, when the first Europeans arrived, it was all over. The population was reduced to about 2,000, living in poverty amidst the ruins of their former culture. The palm tree and several other species had become extinct, leaving the island with only one small species of tree and two species of shrubs.

The story of Easter Island provides a model for our whole planet. The islanders carried out for us the experiment of permitting unrestricted population growth, profligate use of resources, destruction of the environment, and boundless confidence in their religion to take care of the future. The result was an ecological disaster leading to a population crash. A crash on a similar scale for the planet would lead to the deaths of

Stone Moai at Akivi (Photo By Yasuda)

about 1.8 billion people. A well respected 'think tank', known as the Club of Rome, twenty years ago attempted to predict the future of the earth in the twenty-first century. It found that if present trends of economic expansion and population growth go unchecked, there will be a rapid decline of resources, together with soaring pollution and population, until resources become nearly exhausted around AD 2020, with a sharp population crash after AD 2050. The most recent update of the Club of Rome's model is even more alarming, indicating that things are getting worse at a rate more rapid than they had predicted. The parallel between the ecological disaster on Easter Island—isolated in the Pacific—and what is happening on planet Earth—isolated in space—is far too close for comfort (Bahn and Flenley, 1992).

References

Bahn, P. G. (1997): Easter Island or (Man–) Eaters Island? *Rapa Nui Journal* 8 (2): 40.

Bahn, P. G. and J. Flenley (1992): *Easter Island, Earth Island*, Thames & Hudson, London & New York.

Flenley, J. R. (1993): The Palaeoecology of Easter Island, and Its Ecological Disaster, in S. R. Fischer, (ed.), *Easter Island Studies. Contributions to the History of Rapanui in Memory of William T. Mulloy*, Oxbow Monograph 32, Oxford, pp. 27–45.

Flenley, J. R., A. S. M. King, J .T. Teller, M. E. Prentice, J. Jackson, and C. Chew, (1991). The Late Quaternary Vegetational and Climatic History of Easter Island, *Journal of Quaternary Science*, 6: 85–115.

Pavel, P. (1995): Reconstruction of The Transport of the *Moai* Statues and *Pukao* Hats, *Rapa Nui Journal* 9 (3): 69-72.

Steadman, D. W. (1995): Prehistoric Extinctions of Pacific Island Birds: Biodiversity Meets Zooarchaeology, *Science*, 267: 1123–31.

Part II

FORESTS AND ANIMISM

CHAPTER 5

Crocodile, Serpent, and Shark: Powerful Animals in Olmec and Maya Art, Belief, and Ritual

ROSEMARY A. JOYCE

The tropical rain forest was the natural world within which the Maya and Gulf Coast Olmec civilisations developed. Here the rhythm of the annual cycle of rainy and dry seasons, and its effects on tropical rivers and lakes determined the way agriculture, hunting, and fishing were conducted. It is not therefore surprising that images of forest plants and animals are part of the iconography of these cultures. An exploration of the significance of crocodilian, serpent, and shark imagery (often intermingled) in Olmec and Maya art provides a means of approaching the beliefs that these people may have had about the forest, and the ways that they may have enacted those beliefs in rituals.

The Olmec Universe

The earliest imagery from the tropical forest civilisations of Central America is that of the Early and Middle Formative Periods of the Olmec culture that developed on the Gulf coast of Mexico between ca. 1100 and 400 BC (Drucker, 1952; Drucker et al., 1959; Coe and Diehl, 1980; Grove, 1981a; Rust and Sharer, 1988). Here, in major centres filled with monumental earthen buildings and large-scale carved stone sculptures (de la Fuente, 1973, 1981), a series of visual symbols abstracting elements of the natural environment were developed. Specific ritual practices, including blood-letting, mortuary ceremonies, and actions taken to dedicate public spaces, drew on this symbolism to express ideas about the relationship of human beings to the forces of nature.

This iconography and the religious practices were widely shared by other contemporary cultures, from the dry highlands of central Mexico to the Pacific coastal plains of El Salvador (Flannery, 1976b; Grove, 1984, 1987; Marcus, 1989; Sharer, 1989; Joyce, 1992b). Many unrelated languages from this area also share words for basic concepts that have been attributed to the influence of the Gulf Coast Olmec centres (Campbell and Kauffman, 1976). While a number of explanations for this influence, including trade, conquest, and migration, have been suggested, the shared set of beliefs and means of expressing them in art, ritual, and political ceremony are more suggestive of the spread of a religion (Drennan, 1976; Flannery, 1976a; Marcus, 1989; Joyce et al., 1991). In this religion, the tropical forest and its denizens played important roles. Because this religion preceded the rise of later civilisations, like that of the Maya, it shaped the basic attitudes of these later peoples to the forest.

In the Gulf Coast Olmec sites, certain animals of the tropical forest were represented on both monumental stone sculpture and smaller-scale carvings in ceramic, green stone, or obsidian. The presence of a wide range of supernatural animals in Formative Period art is well known. Supernatural animals, or zoomorphs, are beings with features of natural animals, combined with other non-animal or non-natural traits. For example, supernatural animals may wear costume elements, like bracelets.

A widespread, but in the view of many scholars (e.g. Grove, 1973), erroneous, idea about Formative iconography is that many of these representations depict jaguars. In fact, the jaguars (and felines in general) occupy a secondary level of importance. Of more primary importance are the supernatural animals, which seem to stand for major regions of a tripartite division of the universe.

The crocodilian is foremost among depicted animals and was possibly based on the caiman of the rivers of the Gulf Coast (Joralemon, 1976; Pyne, 1976; Stocker et al., 1980; Grove, 1984; Joyce et al., 1991; Reilly, 1991, 1994). Serrated brows, legs, sprouting vegetation, and particular tooth row and eye motifs distinguish this supernatural animal in incised ceramics and stone sculpture.

In its most elaborate, three-dimensional form, represented by a carved stone sarcophagus from the site of La Venta, which enclosed the entire body of a member of the social élite, the crocodilian floats on the surface of water. It is depicted in full body form, where the four legs, pug nose, typical L-shaped eye, and serrated or 'flame' eyebrows are clear (Fig. 1). Each of its paws is decorated with wrist ornaments (probably waterlilies as suggested by Kent Reilly, 1994), and its mouth is closed and rests peacefully on the water. Above the eyes of the crocodilian the so-called 'flame' eyebrows rise in tufts up towards the sky and along its back is a series of vegetation motifs, similarly rising up into the air.

This image of the crocodilian (floating on water and framed in growing plants) demonstrates a close association between this animal and the earth's surface. Some arguments have been offered for a sky identification as well, based on the serrated brow motif and the common occurrence of diagonal crossed bands in the eyes or mouth, a sign commonly used in later Maya art to represent locations in the sky (see Joralemon, 1976). In addition, the serrated eyebrows have been related to the flaming sun, but they may more plausibly be interpreted as sprouting vegetation motifs (which replace other vegetation motifs in highly schematic renderings).

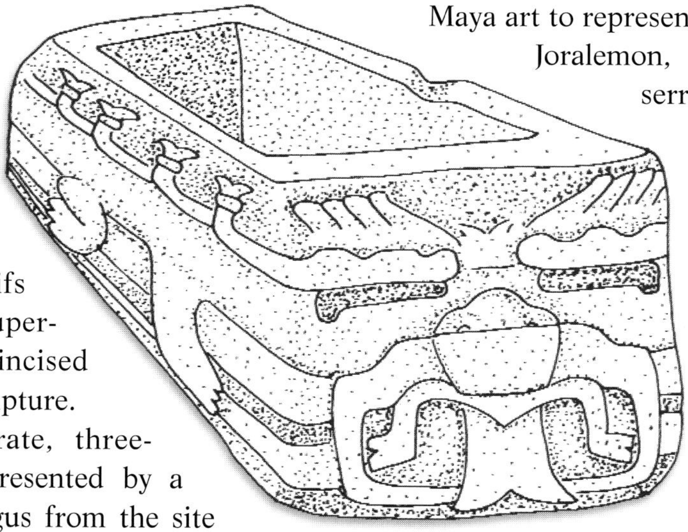

Other Formative Period images use the mouth of the crocodilian, always without teeth, to frame scenes of human beings, perhaps rulers, emerging from an interior space. In scenes carved on the side of a hill at the site of Chalcatzingo, Morelos, and those painted over the mouth of a cave at Oxtotitlan, Guerrero, the frontal view of the crocodilian's mouth is clearly depicted as a cave opening, offering entry into the earth (Grove, 1970, 1973, 1984, 1987).

Fig. 1: Gulf Coast Olmec monument portraying the crocodilian. Note the L-shaped eyes and serrated eyebrows on the front, and the open mouth without teeth. The legs are shown on the side, and waterlilies grow from the back of the crocodilian. Sarcophagus from La Venta, Mound A-3. Middle Formative Period. Drawing by David C. Grove.

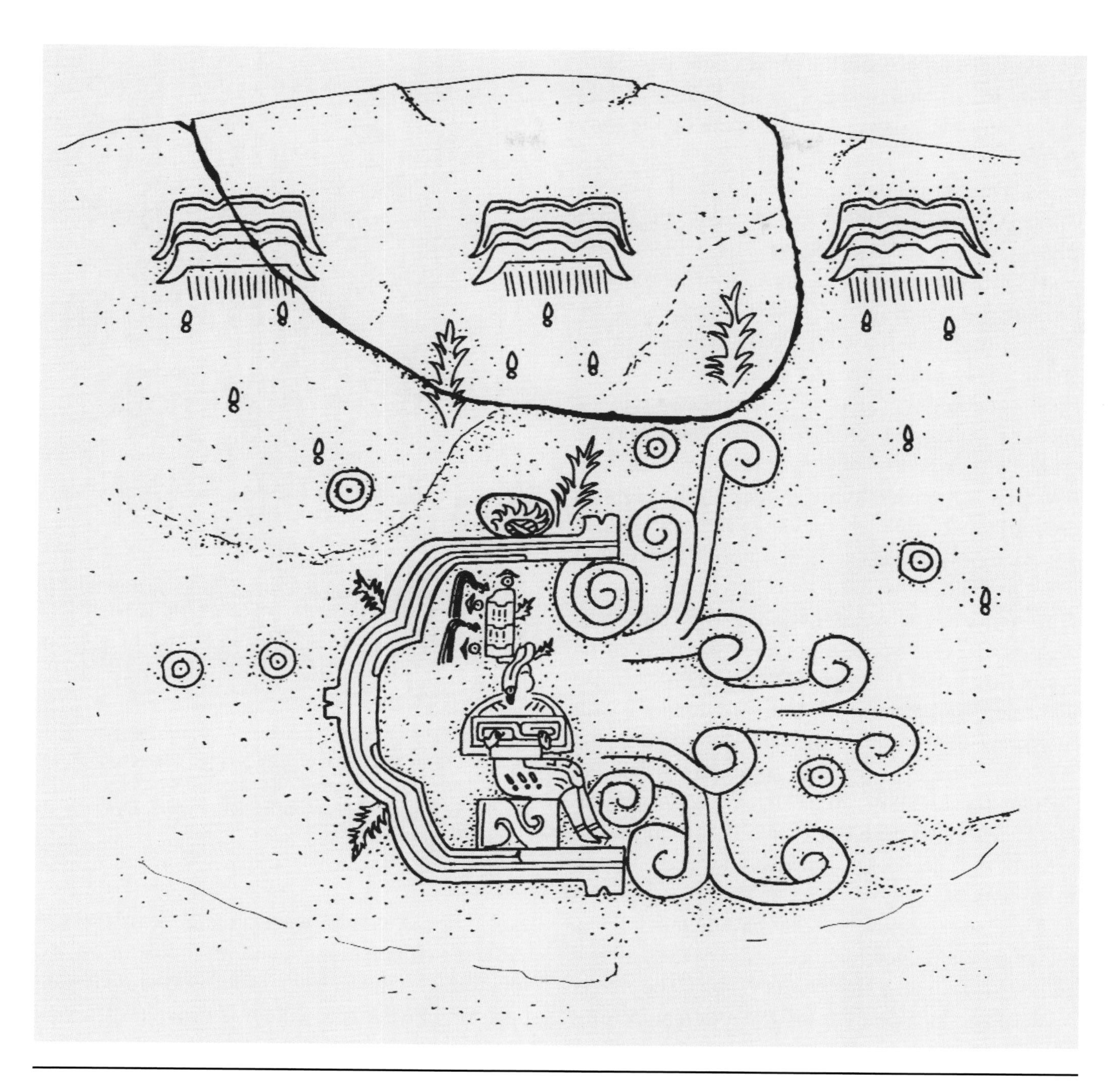

Fig. 2: Seated figure in the profile jaws of supernatural animal. Relief 1, Chalcatzingo. Middle Formative Period. Drawing by David C. Grove.

At Chalcatzingo, the famous Relief 1 shows a seated figure inside the jaws (depicted in profile) of a supernatural animal whose eyes are marked with crossed bands (Fig. 2). Mist comes from the open mouth and clouds and raindrops are shown in front of it. The Chalcatzingo reliefs show plants (perhaps epiphytes which grow on the nearby hill) sprouting around the crocodilian mouth in response to the falling rain drops that descend from clouds in the sky, clouds that are identical in form to the mist which rises from the mouth itself. Separate reliefs along the adjoining rock face depict a sequence of rainfall and the growth of squash plants. The whole composition forms an explicit outline of the model of the earth, personified in animal form, as the source of rain and vegetation, with a human actor at the very centre.

As Donald Lathrap (1973) argued for the caiman carved on the Obelisk Tello of Peru's Chavin culture, these Olmec crocodilians may represent the power of agricultural renewal through the watering of the earth's surface. In the tropical forest of the Gulf Coast this renewal of fertility was ultimately the result of the annual floods of the rivers that were inhabited by living caimans. Like the Obelisk Tello caimans, the Olmec crocodilian's body gives rise to plants. Also like the Obelisk Tello caimans, the Olmec crocodilians are, when viewed from below, provided with organs of generation. This view of the crocodilian is rare in Olmec art, but is represented by a series of pavements buried to create the sacred space of an enclosed plaza at La Venta (Joyce 1987; compare Reilly, 1994, who also identifies the pavements with crocodilians but interprets this motif as the spine of the animal) (Fig. 3). These pavements, often described as 'jaguar masks', are constructed of serpentine blocks arranged on a clay backing, leaving some areas of the clay backing uncovered to form motifs in reserve.

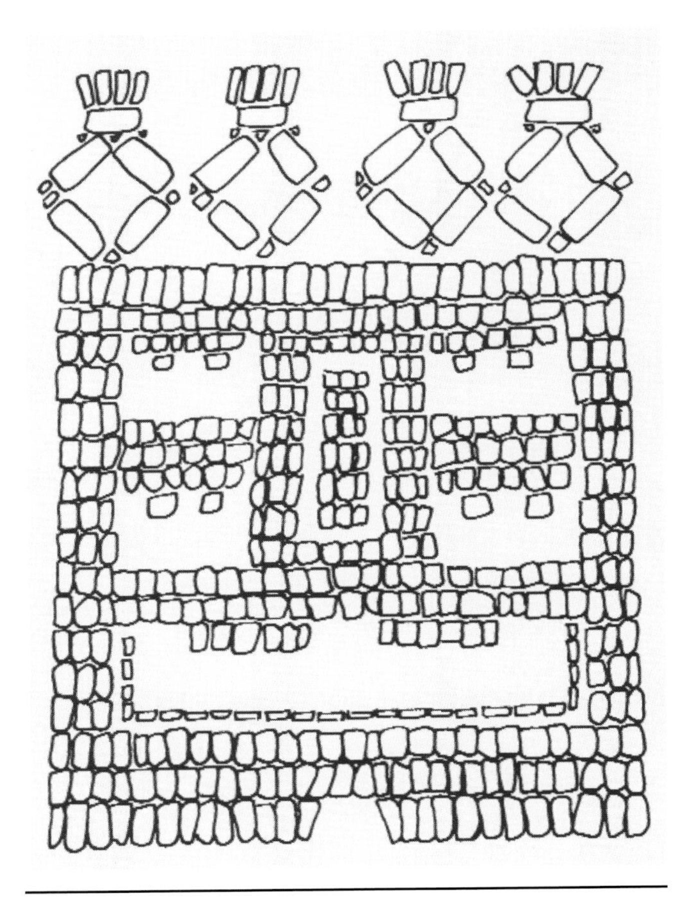

Fig. 3: Schematic image interpreted as crocodilian. Mosaic pavement made of serpentine blocks from La Venta. Middle Formative Period. Drawing by David C. Grove.

The basic outline of the pavements is of a rectangle with a notch at one end. On the notched end, a long area of clay is left visible. The notch, or cleft, and long opening are the minimal features of the frontal face of the crocodilian. Below these features, the pavement has five motifs formed by leaving the clay visible between blocks of stone—four identical elements forming a box around a unique central element. The four repeated motifs have three projections extending towards the crocodilian mouth, which represent the paws and claws of the animal. The central element is a simple rectangle extending along the length of the centre-body, and its most obvious natural analogue is the genitalia of the crocodilian.

Where the Obelisk Tello shows two different sexes, the La Venta pavements repeat a single motif, which is ambiguous as an indicator of sex. Nonetheless, the interpretation presented here suggests that the crocodilian is capable of sexual reproduction, and perhaps associates the fertility of the animal with the general fertility of the land. In support of this argument, the same series of deposits contain other pavements that represent a general plant or world-tree motif—a cross-shape with a mirror at the base (Carlson, 1981; Joyce, 1987). These plant motifs are represented in various stages of growth, from a single trunk to a full cross, and symbolically represent the cycle of regeneration of vegetation.

The Formative Period Mesoamerican crocodilian thus represented the earth's surface, floating on the water, annually renewed by the floods fuelled by the rains of the tropical forest, and supporting the cyclical growth of vegetation. The mouth of the crocodilian was presented as a means of entry into the earth (the source of fertility), and Formative Period rulers depicted themselves inside, or emerging from, this mouth in order to connect themselves with the power of earthly fertility, perhaps especially the cycle of rainfall. A niche carved on the front of so-called altars (probably thrones) which correspond to the open mouth of the earth crocodilian at Chalcatzingo, may also provide evidence of sprouting vegetation motifs as they have a slightly rectangular, cleft shape which has been identified as a corn cob. The ledge of altars is marked with symbols that depict the gums of the crocodilian. The background of Middle Formative bas relief stelae at La Venta also depict these geometric cave mouth forms. Thus for Mesoamerica's earliest civilisation the earth was powerful and animate but not threatening: the teeth of real crocodilians were not shown in the images of their supernatural patron.

In contrast, the second major animal icon of the Gulf Coast Olmec was distinguished by an emphasis on its teeth (Joralemon, 1976; Joyce et al., 1991). In its most elaborate form this animal is clearly recognisable as a monstrous shark. Unlike the crocodilian it has no feet. Its eyes have no pupils or irises and appear to be empty sockets. Its upper jaw is extremely elongated and supports a row of huge, jagged teeth. This second supernatural animal is found in a complementary distribution with, or in paired opposition to, the crocodilian, and was merged with it in Joralemon's (1976) original definition of the Olmec dragon. Crossed bands mark only the body of this second supernatural animal, and must refer to some quality shared by these two creatures which I suggest is their joint representation as the surface of the world, as land surrounded by water.

Shark's teeth themselves were known and used as ornaments by people of the Formative Period, and may have been employed as real or symbolic instruments in bloodletting rituals (Joyce, 1987; Joyce et al., 1991). In contrast to the crocodilian, the shark represents the encircling ocean of salty water, unconnected with the positive benefits of rainfall and the growth of plants. It appears as a symbol of, and an instrument for, the powerful ritual action of personal blood sacrifice, a probable representation of the wild forces outside the world that humans understood and controlled.

In inland areas, away from the coast where sharks could actually be seen, the counterpart to the beneficent crocodilian of the earth was often reinterpreted as a serpentine monster. In the Chalcatzingo reliefs, a writhing serpent's body is attached to a mouth filled with huge shark's teeth (Grove, 1984, 1987). Unlike the standard depiction of the shark, the eye of this monster has a rounded pupil. In its mouth, this shark–snake holds the outline of a struggling human figure, an inversion of the calm serenity of the seated figures depicted seated in the crocodilian mouth at the same site (Fig. 4). In other Formative Period images, human figures grasp or nestle in the folds of serpent–sharks, in attitudes that suggest they are participating in ritual actions.

Fig. 4: Serpent with shark's teeth in mouth. Monument 5, Chalcatzingo. Middle Formative Period. Drawing by David C. Grove.

These two contrasting supernatural animals seem to have been the most important part of the Formative Mesoamerican cosmological model. A third supernatural animal, a raptorial bird that appears more rarely, presumably has celestial connotations. The curved beak of this bird combines with the hand/paw/wing motif and serrated brows, both of which are features shared with the crocodilian. These two creatures (avian and crocodilian) may contrast as a pair with the legless ocean zoomorph, the shark monster. Together they establish a central axis which associates the earth and sky, and thus form a vertical dimension.

These animal symbols were deployed in rituals. Incised ceramic vessels which contrast crocodilian and shark are found in San Lorenzo, Tlatilco, Oaxaca, and even Honduras, and were probably used in public ceremonies. At San Lorenzo, flat bottom, flaring wall bowls with incised geometric motifs depict an abstract crocodilian, shark and, rarely, avian supernatural animals. There appears to be a contrast between crocodilian motifs (found on thickened rim bowls) and shark and avian motifs (found on direct rim bowls). The same supernatural animals are depicted on bowls and bottles from the Central Mexican Highlands. These depictions are often less abstract, and may contrast open forms (associated with crocodilian imagery) and closed ones (associated with shark and avian imagery). Ceramic vessels from Copan are like those of the Mexican Highlands in the range of vessel forms (bowls and bottles), but the supernatural animals are more schematic. The association of sharks with bloodletting, and crocodilians with sprouting vegetation, suggests the possibility that these vessels were used in contrasting rituals.

In Oaxaca, on contemporary incised bowls, another dimension of contrast was employed, namely the frontal and profile views of supernatural animals, the former called the fire serpent and the latter the jaguar. The features of the fire serpent are similar to the profile of the crocodilian image, while the jaguar (marked primarily by a central cleft) conforms to a frontal view of the same animal. In fact, a single vessel from Tlatilco in the Central Mexican Highlands juxtaposes the profile and frontal views, both of which represent the crocodilian. Interestingly, a difference in distribution was noted within San Jose Mogote, Oaxaca, with profile views distributed east and west and frontal views distributed south (with no data on the north).

If these motifs are conceived of as different views of a single supernatural animal then there was a conceptual crocodilian depicted in the centre of San Jose Mogote with its head pointed downwards, towards the south. This depiction recalls the image of the crocodilian as a tree which is found in later Mesoamerican art (Fig. 5). The use of distinct vessels decorated with icons of different supernatural animals in burials at the site suggests one possible context for ceremonies that might have varied in the supernatural patrons being invoked.

The elaborate incised designs on ceramics are,

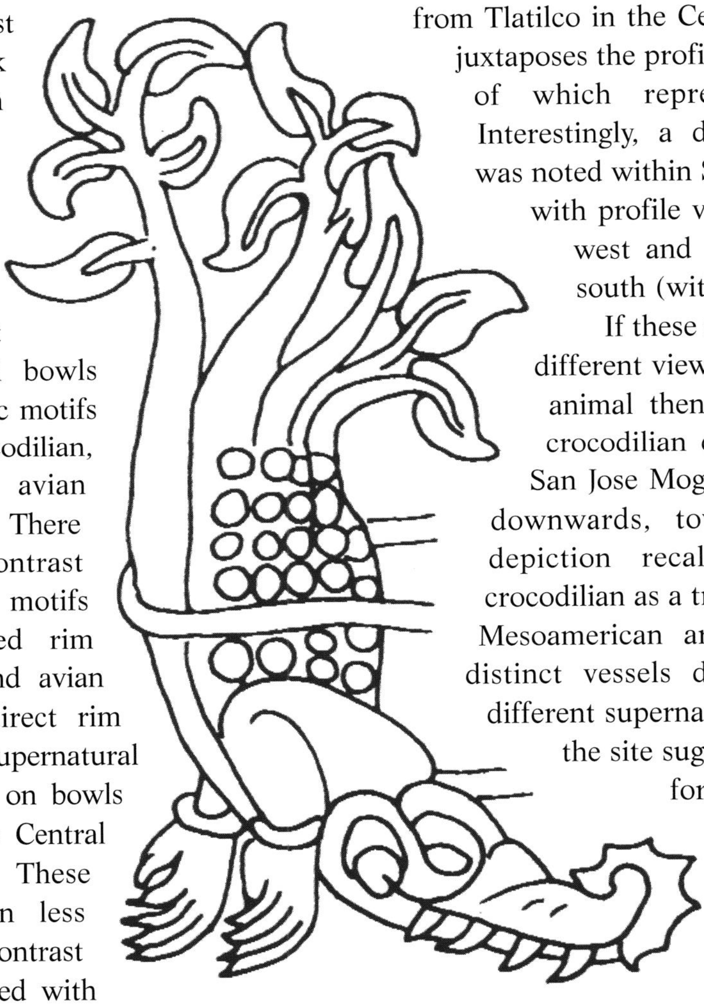

Fig. 5: Crocodilian shown as tree, with snout and front paws as roots and back limbs and tail sprouting vegetation. Izapa Stela 5. Late Formative Period. Drawing by Virginia G. Smith.

for the most part, replaced in the Middle Formative Period by simpler incised decoration. The double line break motif on the rims of dishes has been interpreted by Grove (pers. comm.) as the crocodilian jaw motif, denoting the dish opening as the mouth of a crocodilian or earth-monster. Serrated outline circles (one of the body markings of the shark) are often found depicted on the interior bases of these dishes, and perhaps mark them as receptacles for liquids such as seawater or blood.

After 900 BC a new medium of carved green stones, especially jade, was introduced and used to represent these supernatural animals. Presumably, access to greenstone was restricted more than access to ceramics had been, and ceramics with less complex designs continued to be made and presumably used in a fashion similar to earlier examples. Unlike the earlier ceramic medium, carved stone items were parts of costumes or effigies of regalia depicted as being carried in the hands of anthropomorphic figures. While ceramic vessels were presumably used in ceremonies, including those associated with burial, incised stone items may have been worn or carried as insignia.

In an examination of burials or burial-like deposits excavated at La Venta, I discerned the presence of two distinct costumes (Joyce, 1987). One, with a double-strand jade bead belt and earspools with pendant beads, often in the form of fangs, was worn by individuals buried with stingray spines, jade awls, or shark's teeth, all of which are symbolic and real bloodletting instruments. These individuals often had a serrated pectoral ornament, which recalled the body markings of a shark. The second burial costume included plain ear spools and a distinctive green stone pendant—an effigy of a bivalve shell—and these burials lacked bloodletting instruments. In at least one case, an example of each costume was included in a single tomb, reinforcing the identification of the two as paired opposites.

Once identified in the La Venta sample, these costumes were also distinguishable in samples from other areas. For example, at Chalcatzingo, a pair of burials in a public structure contained the only individuals recovered who wore elaborate greenstone costume elements (Grove, 1984, 1987). One of the two had the diagnostic features of the bloodletting costume, namely the double strand belt and earrings with pendant beads. The second burial, while lacking the bivalve shell pendant, had none of the features of the first costume.

Because the greenstone items represented costume elements, I was able to survey large-scale sculpture looking for comparable costumes. Elements associated with the head and neck proved to be distinctive. Ear spools with pendant beads were present on monumental stone heads and in low reliefs on the Gulf Coast, at Chalcatzingo, and, in at least one case, were associated with a distinctive serrated outline pectoral ornament, like those found with some examples of this costume. In contrast, the bivalve shell pendant was not depicted in any recognisable form. I suggest that the costume dichotomy might correspond to a major dichotomy in the population, such as that between males and females. With few exceptions only males are shown in the monumental art, and all the figures identified wearing earspools with pendant beads were apparently male. The correlation of a lack of female figures and a lack of figures wearing bivalve shells in stone sculpture may support the identification of bivalve shell motifs with women. The single costume element shared by both categories of costumes provides positive evidence to reinforce this negative finding: large discs believed to represent mirrors are shown on male figures in low relief but similar motifs, including actual pieces of iron ore, occur as pectoral pendants on ceramic figurines, among them some clearly female figurines from Tlatilco.

Consequently, it appears that although both

males and females had access to status roles, as signified by the costumes in which they were buried, only males and their costumes were generally depicted in large-scale public art. The role associated with this tentative male costume, which features a serrated pendant that recalls the body markings of the supernatural shark, involves the use of bloodletting instruments. Other costume ornaments worn by these individuals are fang or jaw pendants from earspools, and small skull pendants. This imagery is clearly associated with death and the partition of the body, aspects of the supernatural shark with its exaggerated teeth.

The sole representational item associated with the possible female costume is an effigy bivalve shell pendant. The lack of bloodletting imagery is notable. Bivalve shell pendants hanging from the belt are diagnostic of later Classic Maya women's costume, and distinguishes them even from males dressed in female robes. Further afield, in the Chavin art of South America (which arguably derives from related roots), a male/female complementary dyad is symbolised by univalve (*Strombus*) and bivalve shells (Lathrap, 1973).

Both males and females could adopt a mirror pendant, which presumably represented a third role not limited by gender. The mirror is one of several artifacts employed by shamans cross-culturally, including those in Mesoamerica. Grove (1981b) has argued that Formative Period élites were shamans who based their claim to access to the supernatural (through the cave mouth) on power which was lodged in the so-called 'altars' and portrait sculptures. The élites may thus be viewed as having monopolised shamanism like the early kings in China (Chang, 1983).

Shamanism is often open to both genders, as is the case in modern Mesoamerica. An analysis of a large suite of burials from Tlatilco supports this suggestion. Items such as mirrors, whistles, and masks, which may represent shamanic regalia, and ceramic figurines of dogs, and figurines wearing mirrors or other costume elements believed to represent shamanic costume, are found with both male and female interments. Like the female figurines with mirrors from Tlatilco, these burials suggest that mirror use, which was perhaps shamanic, was initially open to males and females. This form of ritual practice was symbolically associated with the ability to cross into the underworld through the open mouth of the crocodilian earth.

Serpents in Maya Culture

The Classic Maya city–states (ca. AD 250–850) occupied a tropical forest environment similar to that of the Gulf Coast Olmec settlements (Marcus, 1982; Sanders and Webster, 1988; Rice, 1993). On the eastern and western edges of the Maya Lowlands, major tropical rivers experienced an annual cycle of rainfall and flooding that renewed the soil. In these regions, some Maya built systems of canals and raised fields that extended the benefits of annual renewal of the soil away from the rivers into what had formerly been swampy forest edges. In the central Maya Lowlands, surface water drained through porous limestone bedrock that did not allow the formation of rivers and streams. Here, where some of the largest Maya cities, such as Tikal, developed, the capture of rainfall in reservoirs of various types was a crucial part of farming (Scarborough, 1983).

In both riverine and non-riverine parts of the Maya Lowlands, the surface of the earth formed a patchwork of fields and waterways. The Maya, drawing on ancient imagery, depicted this patchwork as the skin on the back of the earth crocodilian, with plants rising from its feet (Fig. 6). The head of this crocodilian was used to represent the earth supporting the base of the great world tree, which bore the sun in a bowl on its head (Schele, 1976; Freidel, 1985; Schele and Miller, 1986; Helmuth, 1988, Schele and Freidel, 1991). As had been the case with Formative Period cultures, the Maya represented the

Fig. 6: Classic Maya crocodilian showing plant motif on back. Copan altar. Late Classic Period. After Maudslay (1889-1902) vol. I, Plate 114.

crocodilian without threatening features (Taylor, 1979; Stone, 1985; Helmuth, 1988). For them, it represented the forest transformed into a supportive home for humans.

In Maya culture, the image of the crocodilian was not placed in opposition to that of the shark as it was in the Formative Period. While the shark is important in Maya art as a counterpart to the jaguar god (Cohodas, 1991) and as a symbolic element distinguishing women's costume (Miller, 1974; Joyce, 1992a), it contrasts with the crocodilian primarily as periphery to centre. For the Maya, the dangerous, challenging realm outside the familiar world was not represented by the ocean (associated with the shark), but instead by supernatural planes of existence reached through ritual action. Access to these other worlds was symbolised by animal imagery.

However, unlike the people of the Formative Period for whom the crocodilian was the primary animal icon of access to the supernatural, for the Maya the preferred animal patron was the serpent. With its long history as a symbol of the wild forces of the natural world in contrast to the predictable aspects of nature embodied in the earthly crocodilian, the serpent imagery represents access to the supernatural as more threatening and dangerous than was the case in the Formative Period.

Serpents were particularly prominent as images of what Susan Gillespie (1993) calls 'pathways' between different planes of existence (see also Schele and Miller, 1986; Bassie-Sweet, 1991). Jewelled serpents formed the branches of the great world tree which was placed in the north to support the sky at the beginning of the contemporary era of Maya cosmogonic time. Double-headed serpents were held as ritual instruments by Maya rulers (Spinden, 1913; Quirarte, 1977) and served as conduits for entry into the world of the living by supernatural beings (Schele, 1985). Serpents were also the embodiment of visions seen by royal women as a result of their ritual sacrifices and were depicted as rising up to emit the spirits of ancestors and gods from their open mouths (Stuart, 1984, 1988; Houston and Stuart, 1989). In these scenes, the serpents display their fangs, and the supernatural beings which emerge from their mouths are thus placed in a peril that reinforces their power, and the power of those who cause them to travel through the serpent–channel (Fig. 7). Perhaps reflecting their ability to control the powerful forces represented in these scenes, one of the insignia of power held by human rulers was a manikin with a snake extending from its leg or genitalia (Fig. 8).

Classic Maya images suggest a continuity of

Fig. 7: Classic Maya serpent as conduit of vision. Yaxchilan Lintel 17. Late Classic Period. Drawing by Ian Graham.

the ancient concepts of the personification of the earth as a crocodilian floating in the primordial sea, and of serpentine monsters as images of power that humans can control from beyond the limits of domesticated space (compare Quirarte, 1977). The crocodilian earth, with its cave-mouth became the frame for human action, the surface on which movement takes place (Freidel, 1985). It is however the mouth of the serpent that is depicted around the doorways of temples (Schavelzon, 1980). Despite the formal similarity with Olmec images of humans seated in the crocodilian mouth, and the implied presence in Maya temples of participants in ritual, the actors

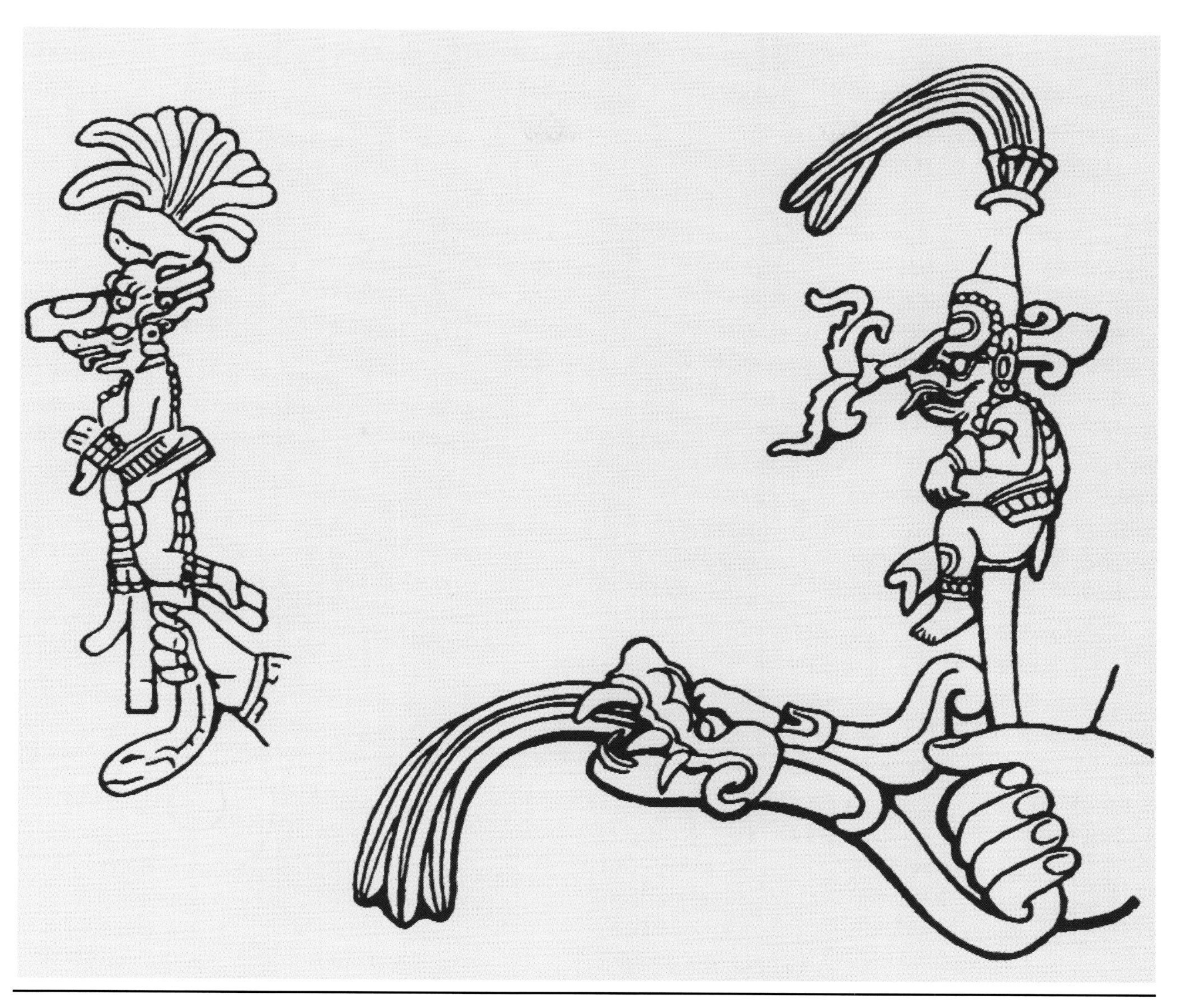

Fig. 8: Classic Maya manikin sceptres depicting serpent as foot. Details from Yaxchilan Lintel I (left) and Quirigua Altar P (right). Late Classic Period. Illustration by C. C. Willoughby.

engaged in rituals taking place in Maya temples must be identified with the supernatural beings that were depicted as travelling through the serpent from beyond the limits of everyday space.

In the books created by the descendants of the Classic Maya following the introduction of European script to the Yucatan peninsula, the same three animals—serpent, shark, and crocodile—are invoked and recalled as beings associated with the forces of creation. Here, as in the earliest images from the forest civilisation of the Olmec, the shark and the crocodilian are paired as creators: 'Who created him? He was the creation of Chac Uayab Xoc (the Great Supernatural Shark), Chac Mumul Ain (the Great Supernatural Crocodile)' says a chant from a healer's handbook, the *Rituals of the Bacabs* (Roys, 1965). This statement echoes other accounts of creation recorded in the prophetic histories, the Books of Chilam Balam (e.g. see Edmonson, 1982). The *Rituals of the Bacab* describe the serpent's role at the beginning of creation, likening the snake to shoots of a tree that was planted by the Creators.

The same concepts are depicted in scenes from the few late Maya manuscripts that survived the Spanish conquest (Tozzer and Allen, 1910). In the New Year ceremonies depicted in the Codex Dresden (Thompson, 1972), the ritual recreating the four world directions involves the placing of an upright image: a tree dressed in a cape and loincloth. In the branches of the tree a serpent is coiled, ready to move between the different planes of existence that were separated at creation when the sky was raised up from the earth's surface and the world's trees were put in place between the earth and the sky. In the pages of the codex, gods are shown emerging from serpents, which are therefore conduits between the supernatural and natural worlds. Counterposed to this vertical axis, the (rarer) depictions of the crocodilian show it as a horizontal surface, in one form as the horizontal plane above (from which falls the rain) and in another form as the horizontal plane below, which together frame the actions of the supernatural beings and ritual impersonators.

From the earliest sculptures of the rain forest civilisation of the Olmec, to the latest words and drawings of the lowland Maya, Mesoamerican cultures used images of these powerful reptilian animals as symbols of the earth and the realms surrounding it. The imagery they chose depended on their observations of the behaviour of the natural animals on which these supernatural creatures were modelled. The imagery embodied these peoples' understanding of the patterns of development of the tropical forest in which they lived.

References

Bassie-Sweet, K. (1991): *From the Mouth of the Dark Cave*, University of Oklahoma Press, Norman.

Campbell, L. and T. Kauffman (1976): A Linguistic Look at the Olmecs, *American Antiquity,* 41 (1): 80–9.

Carlson, J. (1981): Olmec Concave Iron-Ore Mirrors: The Aesthetics of a Lithic Technology and the Lord of the Mirror (with an Illustrated Catalogue of Mirrors), in E. Benson, (ed.), *The Olmec and Their Neighbors*, Dumbarton Oaks, Washington, pp. 117–48.

Chang, K. C. (1983): *Art, Myth and Ritual: The Path to Political Authority in Ancient China*, Harvard University Press, Cambridge, MA.

Coe, M. D. and R. Diehl (1980): *In the Land of the Olmec, Volume I: The Archaeology of San Lorenzo Tenochtitlan*, University of Texas Press, Austin.

Cohodas, M. (1991): Ballgame Imagery of the Maya Lowlands: History and Iconography, in V. Scarborough, and D. Wilcox, (eds.), *The Mesoamerican Ballgame*, University of Arizona Press, Tucson, pp. 251–88.

De la Fuente, B. (1973): *Escultura monumental olmeca: catálogo. Cuadernos de historia del arte, I*, Instituto de Investigaciones Esteticas, Universidad Nacional Autonoma de Mexico, Mexico D.F.

—— (1981): Toward a Conception of Monumental Olmec Art, in E. Benson, (ed.), *The Olmec and Their Neighbors*, Dumbarton Oaks, Washington, pp. 83–94.

Drennan, R. (1976): Ritual and Ceremonial Development at the Early Village Level, in K. Flannery, and J. Marcus, (eds.), *The Cloud People*, Academic Press, New York.

Drucker, P. (1952): *La Venta, Tabasco: A Study of Olmec Ceramics and Art*, Smithsonian Institution, Bureau of American Ethnology Bulletin 153, Washington, DC.

Drucker, P., R. Heizer, and R. Squier (1959): *Excavations at La Venta, Tabasco, 1955*, Smithsonian Institution, Bureau of American Ethnology Bulletin 170, Washington, DC.

Edmonson, M. (trans.) (1986): *The Ancient Future of the Itza: The Book of Chilam Balam of Tizimin*, University of Texas, Austin.

Flannery, K.V. (1976a): Contextual Analysis of Ritual Paraphernalia from Formative Oaxaca, in K. Flannery, (ed.), *The Early Mesoamerican Village*, Academic Press, New York, pp. 333–45.

Flannery, K. (ed.), (1976b): *The Early Mesoamerican Village*, Academic Press, New York.

Friedel, D. (1985): Polychrome Façades of the Lowland Maya Preclassic, in E. Boone, (ed.), *Painted Architecture and Polychrome Monumental Sculpture in Mesoamerica*, Dumbarton Oaks, Washington, pp. 5–30.

Gillespie, S. (1993): Power, Pathways and Appropriations in Mesoamerican Art, in D. and N. Whitten, (eds.), *Imagery and Creativity: Ethnoaesthetics and Art Worlds in the Americas*, University of Arizona Press, Tucson, pp. 67–107.

Grove, D. C. (1970): *The Olmec Paintings of Oxtotitlan Cave, Guerrero, Mexico*. Studies in Pre-Columbian Art and Archaeology, No. 6, Dumbarton Oaks, Washington.

—— (1973): Olmec Altars and Myths, *Archaeology,* 26 (2):128–35.

—— (1981a): *The Formative Period and the Evolution of Complex Culture. Handbook of Middle American Indians, Supplement I*, University of Texas Press, Austin, pp. 373–91.

—— (1981b): Olmec Monuments: Mutilation as a Clue to

Meaning, in E. Benson, (ed.), *The Olmec and Their Neighbors*, Dumbarton Oaks, Washington, pp. 49–68.

—— (1984): *Chalcatzingo: Excavations on the Olmec Frontier*, Thames and Hudson, London.

—— (ed.). (1987): *Ancient Chalcatzingo: The People of the Cerros*. University of Texas Press, Austin.

Helmuth, N. (1988): *The Surface of the Underwaterworld*, Doctoral Dissertation, University of Hamburg, Germany.

Houston, S. and D. Stuart (1989): *The Way Glyph: Evidence for 'Co-Essences' Among the Classic Maya, Research Reports on Ancient Maya Writing 30*, Center for Maya Research, Washington.

Joralemon, P. D. (1976): The Olmec Dragon: A Study in Precolumbian Iconography, in H. Nicholson, (ed.), *Origins of Religious Art and Iconography in Preclassic Mesoamerica*, UCLA Latin American Center/Ethnic Arts Council of Los Angeles, Los Angeles, pp. 27–72.

Joyce, R. (1987): *Ceremonial Roles and Status in Middle Formative Mesoamerica: The implications of Burials from La Venta, Tabasco, Mexico*, Paper presented at the III Texas Symposium on Mesoamerican Archaeology, Program in Latin American Studies, University of Texas, Austin.

—— (1992a): Dimensiones simbolicas del traje en monumentos clasicos Mayas: La construccion del genero a traves del vestido, in L. Asturias, and D. Fernandez (eds), *La Indumentaria y el Tejido Mayas a Traves del Tiempo*, Monograph 8 of the Museo Ixchel del Traje Indígena, Guatemala, pp. 29–38.

—— (1992b): Innovation, Communication and the Archaeological Record: A Reassessment of Middle Formative Honduras, *Journal of the Steward Anthropological Society*, 20 (1 and 2): 235–56.

Joyce, R., R. Edging, K. Lorenz and S. Gillespie (1991): Olmec Bloodletting: An Iconographic Study, in M. G. Robertson, and V. Fields, (eds.), *Sixth Palenque Round Table, 1986*, The Palenque Round Table Series, Volume 8, University of Oklahoma Press, Norman, pp. 143–50.

Lathrap, D. (1973): Gifts of the Cayman: Some Thoughts on the Subsistence Basis of Chavin, in D. Lathrap, and J. Douglas (eds.), *Variation in Anthropology*, Illinois Archaeological Survey, Urbana, IL, pp. 91–105.

Marcus, J. (1982): The Plant World of the Sixteenth and Seventeenth-century Maya, in K. Flannery (ed.), *Maya Subsistence: Studies in Memory of Dennis E. Puleston*, Academic Press, New York, pp. 239–73.

—— (1989): Zapotec Chiefdoms and the Nature of Formative Religions, in R. Sharer, and D. Grove (eds.), *Regional Perspectives on the Olmec*, Cambridge University Press, Cambridge.

Maudslay, A. P. (1889–1902): *Biologia Centralii-Americana, or Contributions to the Knowledge of the Flora and Fauna of Mexico and Central America. Archaeology*, 4 vols., London.

Miller, J. (1974): Notes on a Stela Pair Probably from Calakmul, Campeche, Mexico, in M. Robertson (ed.), *Primera Mesa Redonda de Palenque, Part I*, The Robert Louis Stevenson School, Pebble Beach, CA, pp. 149–61.

Pyne, N. (1976): The Fire–Serpent and Were–Jaguar in Formative Oaxaca: a Contingency Table Analysis, in K. Flannery (ed.), *The Early Mesoamerican Village*, Academic Press, New York, pp. 272–80.

Quirarte, J. (1977): Early Art Styles of Mesoamerica and Early Classic Maya Art, in R.E.W. Adams (ed.), *The Origins of Maya Civilization*, University of New Mexico Press, Albuquerque, pp. 249–84.

Reilly, F. K., III (1991): Olmec Iconographic Influences on the Symbols of Maya Rulership: An Examination of Possible Sources, in M. G. Robertson, and V. Fields (eds.), *Sixth Palenque Round Table, 1986*, The Palenque Round Table Series, Volume 8, University of Oklahoma Press, Norman, pp. 151–66.

—— (1994): Cosmología, soberanismo y espacio ritual en la Mesoamérica del Formativo, in J. Clark (ed.), *Los olmecas en Mesoamérica*, El Equilibrista & Turner Libros, Mexico and Madrid, pp. 239–59.

Rice, D. (1993): Eighth-century Physical Geography, Environment, and Natural Resources in the Maya Lowlands, in J. Henderson, and J. Sabloff (eds.), *Lowland Maya Civilization in the Eighth Century A.D.*, Dumbarton Oaks, Washington, pp. 11–63.

Roys, R. (trans.) (1965): *Rituals of the Bacabs*, University of Oklahoma Press, Norman.

Rust, W. and R. Sharer (1988): Olmec Settlement Data from La Venta, Tabasco, Mexico, *Science*, 242: 102–104.

Sanders, W. and D. Webster (1988): The Mesoamerican Urban Tradition, *American Anthropologist*, 90: 521–46.

Scarborough, V. (1983): A Late Preclassic Water System, *American Antiquity*, 48: 720–44.

Schavelzon, D. (1980): Temples, Caves or Monsters? Notes on Zoomorphic Facades in Pre-Hispanic Architecture, in M. G. Robertson (ed.), *Third Palenque Round Table, 1978, Part II*, University of Texas Press, Austin, pp. 151–62.

Schele, L. (1976): Accession Iconography of Chan-Bahlum in the Group of the Cross at Palenque, in M. G. Robertson (ed.), *The Art, Iconography and Dynastic History of Palenque, Part III*, Robert Louis Stevenson School, Pebble Beach, CA, pp. 9–34.

—— (1985): The Hauberg Stela: Bloodletting and the Mythos of Maya Rulership, in M. G. Robertson, and V. Fields (eds.), *Fifth Palenque Round Table, Vol. VII*, Pre-Columbian Art Research Institute, San Francisco, pp. 135–49.

Schele, L. and D. Freidel (1991): The Courts of Creation: Ballcourts, Ballgames, and Portals to the Maya Otherworld, in V. Scarborough, and D. Wilcox (eds.), *The Mesoamerican Ballgame*, University of Arizona Press, Tucson, pp. 289–316.

Schele, L. and M. E. Miller (1986): *The Blood of Kings: Dynasty and Ritual in Maya Art*, Kimbell Art Museum, Fort Worth.

Sharer, R. (1989): The Olmec and the Southeast Periphery of Mesoamerica, in R. Sharer, and D. Grove (eds.), *Regional Perspectives on the Olmec*, Cambridge University Press, Cambridge, pp. 247–71.

Spinden, H. (1913): *A Study of Maya Art: Its Subject Matter and Historical Development*, Peabody Museum of Archaeology and Ethnology, Harvard University, Memoirs vol. 6, Cambridge, MA.

Stocker, T., S. Meltzoff, and S. Armsey (1980): Crocodiles and Olmecs: Further Interpretations in Formative Period Iconography, *American Antiquity,* 45: 740–58.

Stone, A. (1985): Variety and Transformations in the Cosmic Monster Theme at Quirigua, Guatemala, in M. G. Robertson, and V. Fields (eds.), *Fifth Palenque Round Table, Vol. VII*, Pre-Columbian Art Research Institute, San Francisco, pp. 39–48.

Stuart, D. (1984): Royal Auto-Sacrifice Among the Maya: A study of Image and Meaning, *Res,* 7/8: 6–20.

—— (1988): Blood Symbolism in Maya Iconography, in E. Benson, and G. Griffin (eds.), *Maya Iconography*, Princeton University Press, Princeton, pp. 175–221.

Taylor, D. (1979): The Cauac Monster, in M. G. Robertson, and D. Jeffers (eds.), *Tercera Mesa Redonda de Palenque,* Vol. 4, Pre-Columbian Art Research Center, Palenque, Mexico, pp. 79–89.

Thompson, J. E. S. (1972): *A Commentary on the Dresden Codex, A Maya Hieroglyphic Book*, The American Philosophical Society, Philadelphia.

Tozzer, A. M. and G. Allen (1910): *Animal Figures in the Maya Codices*, Peabody Museum of Archaeology and Ethnology, Harvard University, Papers Vol. IV, No.3, Cambridge, MA.

CHAPTER 6

The Snake Cult in Japan

HIROKO YOSHINO

Introduction

Many pre-literacy religions throughout the world have snake cults which are usually linked with ancestor worship. One theory suggests that the cult originated in ancient Egypt from where it spread both east and west. To the east, the cult spread through India, the Far East, and the Pacific Islands, and even to the Americas.

In ancient Egypt the cobra symbolised fire and the sun. It adorned the crown and brow of kings and gods. In India too the cobra, or *naga*, was deified and became an object of worship, influencing Hinduism and Buddhism. The shedding of its skin was a symbol of everlasting life, rebirth, and purification (Hastings, 1976).

Snake Worship in Japan

As a part of this diffusion eastward through Asia to the Pacific, Japan acquired its own snake cult. Pre-literate peoples in many parts of the world associated snakes with ancestor worship, and there are many reasons for this association. The most fundamental of these are:

1. the snake's shape which has phallic associations;
2. the way the snake instantly kills its prey with strong poison;
3. the renewal of life through the shedding of the snake's skin.

These characteristics are employed as symbols of the following:

1. male fertility and life;
2. power;
3. immortality (guaranteed through the shedding of the skin).

These symbols then gradually became associated with ancestor worship.

As is known, pottery and clay figurines (*dogu*) of the Middle Jomon period (ca. 4000 BC to 3000 BC) were full of lively and realistic snake figures. Here we find vivid evidence of snake worship. Such figures express a strong belief in the snake as an ancestor-god (Fig. 1). The snake is not just another low-level deity such as the 'water-god', but a more powerful, omnipotent deity, of which the 'water-god' is also one aspect. His real identity is that of an ancestor-god. The pure and vivid

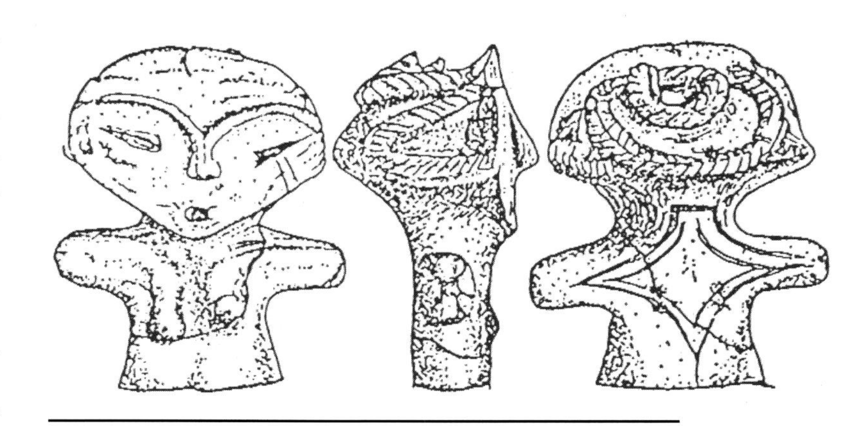

Fig. 1. Clay figurine of Middle Jomon period

form of ancestor-worship changes with the transition to the Yayoi culture period (ca. 400 BC to AD 200) when the snake's aspect as a deity diversifies.

Changes and Development of the Snake Cult

There are no snake figures depicted on the pottery of the Yayoi peoples, who introduced rice agriculture to Japan. People of many origins crossed over to the Japanese islands during the Yayoi period and with them many forms of religion must have come. However, the apparent disappearance of depicted snakes from this period does not mean that the snake cult, with its strong association with ancestor worship, had declined. Rather, it was reinforced by the incorporation of other symbolic elements and through diversication making the cult became even more influential.

By 'diversification' I mean that things such as trees and mountains, perceived to be similar in form to the snake, were invested with the same spiritual meaning, and became objects of worship. The addition of other symbolic elements refers to the perception of the snake god as protector of rice fields and storehouses, and also of cereal crops in general, which derives from the fact that the snake is the natural enemy of the mouse.

Changes in subsistence patterns and accompanying intellectual and emotional changes led to differences in the way in which the snake cult was expressed. That of the Middle Jomon period can be described as direct, simple, and naive, and that of Yayoi period subtle, complex, and sophisticated.

The Yayoi people in all probability had various different cultures of their own, whether from China or elsewhere. Some of these cultures must have had their own snake cults. The Yayoi people had other symbols, such as trees and mountains, which they worshipped as symbols of the snake, rather than the snake itself. The following is a study of this process.

The Beginning of *Mitate* (Substitution)

■ MOUNTAINS

The ancient Japanese believed that the 'correct' form of the snake was its coiled form. The *koshiki*, the rice-cooking pot of the Yayoi period, resembles an inverted coiled snake, so that the snake, whether real or a representation one made from rope, is formally depicted in the *koshikitate* form. For example, the *ryuhebisama* (dragon-snake), which is the central figure of the Oimi festival of the Izumo shrine, is represented in this pose, and in my view the *kagami-mochi* (rice cake), traditionally offered to the ancestors for the New Year celebration, is a symbol of the snake cult.

If the *kagami-mochi* is the smallest example of such symbolisation then the largest are the mountains. The ancient Japanese likened the conical shape of mountains to the coiled snake–ancestor. Many beautiful conical-shaped mountains in Japan, such as Mt. Miwa in Yamato, Mt Nantai of Nikko, and Mt Haruna have an associated snake–god legend precisely because of this.

■ HOUSES AND STRUCTURES

The conical shaped dwellings used throughout the Jomon and Yayoi periods were also suggestive of coiled snakes to the ancient Japanese. The interior of the dwelling was symbolic of the snake's womb, and was also a sacred place of ritual. In the Suwa area of Shinshu, a winter ritual was conducted in front of a snake figure made of straw. The snake, which was a symbol of the ancestor–gods, was placed in a *tsuchi-muro*, a ceremonial structure called the 'snake house'. Other examples of structures symbolising the snake's womb are the *ubuya* (a structure for childbirth), the *moya* (a structure for mourning), and the *kariya* (temporary structures used in ceremony), or *guro*, *gura*, (ceremonial structures rebuilt for each festival). Of these, the practice of building birthing and funeral structures has died out, but the making of *kariya* is still an important part of many traditional festivals.

▪ TREES AND VINE PLANTS

The snake can be likened to a large stick, with no limbs; phallic in shape. The tree, with a single perpendicular trunk that does not branch out (therefore also phallic in shape) is often regarded as sacred, suggestive of the ancestor–god. If the tree symbolises the snake at rest, the vine plants are suggestive of active, moving snakes. The head of the snake may be represented by the Japanese bladder plant (Chinese lantern plant, *Physalis alkekengi*). In this way, there are many varieties of plants which came to be considered sacred because of their physical resemblance to the snake, although the majority are trees (Table I). In particular, the tree most symbolic of the snake–ancestor–god was the subtropical *hoki* tree (Fig. 2).

In the *utaki* of Okinawa (a sacred area in the forest for the worship of gods), the sacred tree which serves as the focus for the manifestation of the gods is usually a *hoki* tree. Many *utaki* have snake–gods, and it is forbidden to mention the word for snake (*pau*) in the *utaki*. Therefore, since the god of the *utaki* is the snake and the great *hoki* tree is the sacred tree through which the gods focus, it follows that the *hoki* tree is a symbol of the snake–ancestor–god.

The fan (*ohgi*) is an essential item for any Japanese festival (*matsuri*). I believe that the origin of the fan in Japan, which has long been unclear, lies in the *hoki* leaf, and I have documented evidence to support this theory elsewhere (Yoshino, 1986). Through the *hoki*, connections can be seen among the fan, the snake cult, and tree worship.

The Sacred Hoki Tree and the Fan

The *hoki* has many names: its botanical name is *biro*, the ancient Japanese called it ajimasa, in Okinawa it is *kuba* and in Chinese it is *hoki*. It was often confused with the betel palm (*binroh*) which is why their botanical names are so similar. The *hoki* is most common on the islands south of

Table I. Plants which are suggestive of snakes

Passive Snake (phallic)	*hoki*, *shuro* (hemp palm), *binroh* (betel palm)
	Nagi (Japan laurel), *ogatama* (*Magnolia compressa*), *Honoki* (*Magnolia hypoleuca*)
	bamboo, pine, cedar, *hinoki* cypress
Active Snake (snaking)	Japanese wisteria, cane
	climbing plants
Suggestive of snake's head	*hozuki* (Japanese bladder plant)
Other	*kaji* (paper mulberry or tapa) (based on *kaiji* method)

Fig. 2. Photograph of *hoki* tree

Kyushu. It stands straight and tall, and has no branches; its broad, palm-shaped leaves grow out of the top of the tree. The leaves are very fibrous and are suitable for many uses when dried, from thatching roofs and walls of houses, to material for clothing and even water vessels. Unlike the palm tree, the leaves of the *hoki* divide into fronds from the middle of the leaf, so that by cutting off the outer part of the leaf, you can make a fan (Fig. 3). Thus, the *hoki* tree's leaves were an essential item of everyday life. In Okinawa, the phrase '*kupa-nu-ha-yu*' had overtones of nostalgia for past times and lifestyles. However, the exceptional feature of the *hoki* is not its everyday utility but its sanctity.

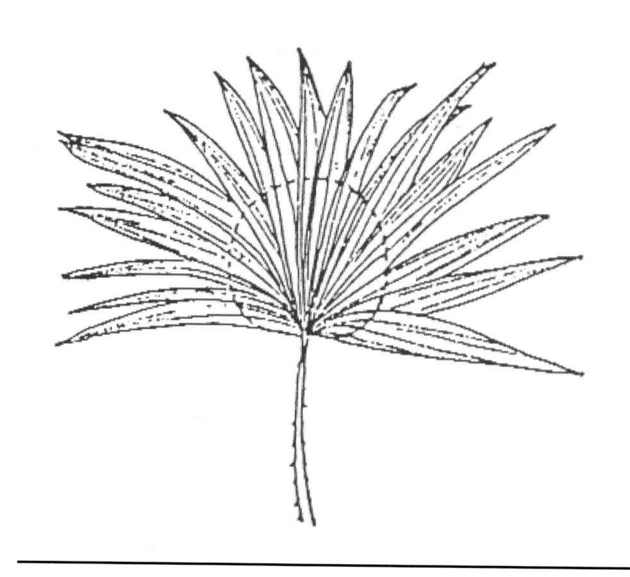

Fig. 3. *Hoki* leaf fan

In Okinawa, gods are worshipped in the *utaki*. Festivals take place in the *utaki*, where the gods may come. A typical *utaki* is a clearing in the forest at the foot of a mountain bordered by a low stone wall. The spot considered to be most sacred, where the sacred tree stands, is called *ibi*, and incense burners are placed there. Although today the sacred tree may be a *fukuki* (*Garcinia spicata*), *deigo* (coral tree, *Leguminosae, Erythrina*) or other species, in the past it was always a *hoki*. Evidence for this can be seen in the *Ryukyu koku yurai ki* (History of Ryukyu) in which there are sixty-nine entries where place names and descriptions of *utaki* are associated with the *hoki*, but not a single entry associated with any other trees. An example that shows the close relationship between the *hoki*, the gods, and *utaki* names is that of the *Kubo utaki* of Kudaka island which has a legend of the Creation God who made the Kingdom of Ryukyu. *Kubo* means *kupa* (i.e. *hoki*) and therefore *Kubo utaki* means *hoki utaki*.

Because the *hoki* was worshipped as the sacred symbol of the *utaki* and was the agent (receptacle) through which the gods manifested themselves, its name was used to identify the forest. The ascension ceremony of the priestess *Kikoe no okimi* was performed at the *Saifuro utaki* (which is equivalent in rank with the Ise shrine in Honshu), and the ceremonial structure built for the holy marriage rite was always made from the *hoki* tree, and the roof and walls were covered with *hoki* leaves. On the mainland, the *hyakushicho*, a ceremonial structure built for the *misogi no gi* (ascension ceremony) of a new emperor was also thatched with *hoki* leaves. Therefore, the *hoki* tree has always played a part in a very important ritual for the most important religious leader in the land.

Use of the *hoki* in a sacred marriage ritual is also recorded. For example, the mute son of the Emperor Suijin, Homutsuwake, and the snake goddess Himenagahime were married in Izumo. The *nihon shoki*, the ceremonial house constructed for the ritual, was named the *Ajimaki no nagaho no miya*, which implies that it must also have been made using *hoki* leaves.

Fuyu Iba (pers. comm.) reports that in Okinawa, the *niidokoro* (the house of the family which establishes a village) is located next to the *utaki*, and plays a leading role in ancestor-worship rituals. Okinawan villages always have a *kamidana*, or altar, where two ancestor-gods are worshipped. The items that serve as receptacles for the gods (the *yorishiro*) are often a pair of *hoki* fans, one of which represents the male god, called the *hi-ohgi* (sun fan), and the other the female god, named the *tsuki-ohgi* (moon fan). The edge of the moon fan is cut off. All the people of the village are said to have descended from these two ancestor-gods.

Thus, the *hoki* tree and leaves are symbols of the ancestor–god and the ancient Japanese perceived the *hoki* to be a receptacle (*yorishiro*) for the ancestor–god. The *yorishiro* is a substitute for, or symbol of, the god, and 'inherits' its sacred nature. The *hoki*'s sacred nature is thus equivalent to that of the god itself. As the gods do not show themselves, the *hoki* is the earthly manifestation of the god. It follows that the form of the *hoki* must be similar to, or is associated with, the form of the ancestor–god itself.

Here, the physical similarities are between the ancestor–god and the *hoki* tree. But the tree itself is rooted in the ground, immovable. In rituals conducted outside the *utaki*, *hoki* leaves were used as *yorishiro*, as substitutes for the tree. If we think of the *hoki* tree as the 'first order' *yorishiro*, then the *hoki* leaf would be the 'second order' and, as I will show in the next section, the fan can be thought of as the 'third order'. In this way, the sacred nature of each *yorishiro* was 'relayed' to the next. This is how the two fans enshrined at the altar of the *niidokoro* became the *yorishiro* of the ancestor–gods.

Fig. 4. Illustration of the entrance to the shrine halls of the Atsuta shrine

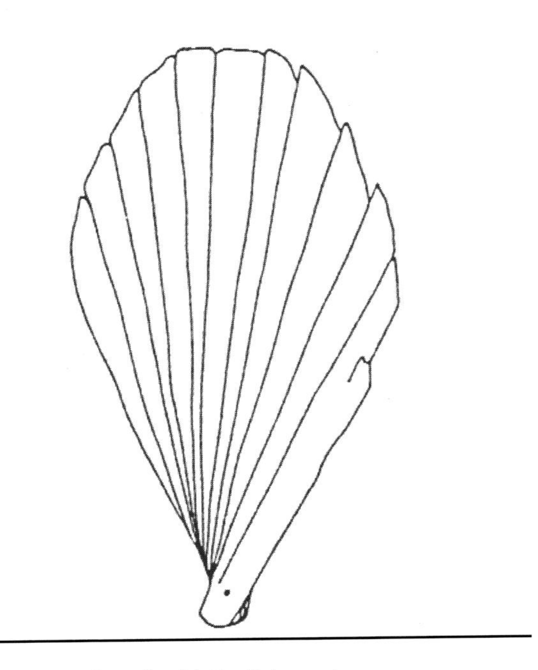

Fig. 5. Fan found at the Heijo Palace site

In mainland Japan there are similar examples of the *hoki* and snake–ancestor–god association, besides the *hinagahime* example. At the entrance to the shrine halls of the Atsuta shrine, where the *kusanagi-no-tsurugi* (sword of Kusanagi), which contained the spirit or soul of the snake spirit Yamatanoorochi, is enshrined, there was a sacred tree thought to be a *hoki* (Fig. 4). In ancient times the vehicles of *daijo tenno* and various princes were roofed with *hoki* leaves called *biroge*. The vehicle can be thought of as a temporary residence, and the same principle as that of the temporary structure in the *taishosai* was applied here.

The fan found at the Heijo Palace site is usually believed to be the original (early) form of the *hiogi* (Japanese cypress fan, Fig. 5), but its shape clearly resembles the *hoki* leaf. During the *Aofushigaki shinji* ritual of Miho shrine in Izumo, the *Toya* (ceremonial leader) who is possessed by the *kami* holds a 'long fan' (*chokei no ogi*, Fig. 6), which looks just like a *hoki* leaf.

The *hokkio,* used by *shugendo yamabushi*

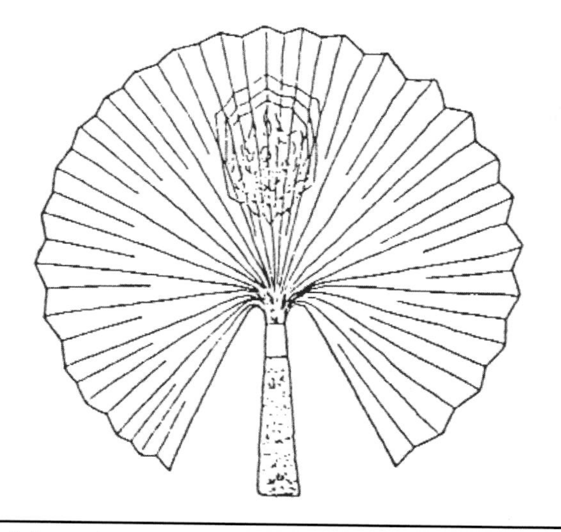

Fig. 6. Long fan held by the *toya* during the *Aofushigaki shinji* ritual of the Miho shrine in Izumo

(ascetics) when entering the mountains, literally uses the word *hoki*. Many different kanji may be used for this fan but the name clearly comes from the *hoki* leaf fan. However, it is diamond shaped derived from the diamond markings often found on poisonous snakes. It is said that hanging this fan from your waist prevents poisonous snake bites. A peacock feather may also be attached to the fan as a further measure, as peacocks are said to eat poisonous snakes.

The fan may also serve as *goshintai* (sacred substance, or *kami*-body) or *shinmon* of shrines with long histories (Sata shrine of Izumo). Famous examples are the Nachi shrine in Kumano, where the 'fan festival' prays for the well-being of the fan god, the *Ogitate matsuri*, of the Kumano shrine, and in the Tohoku area, the *Kurokawa noh* (a play) in which the *ogi-sama*, or fan god, is one of the characters (Fig. 7).

Today, the established theory is that the fan has its value in its *suehirogari* shape which is considered lucky. However, is its 'lucky' shape reason enough for the fan to be worshipped as *goshintai* and as an important item for which festivals are held? It is the association of the fan with the snake–ancestor–god, and the ritual symbolism associated with it, that makes the fan such a key item.

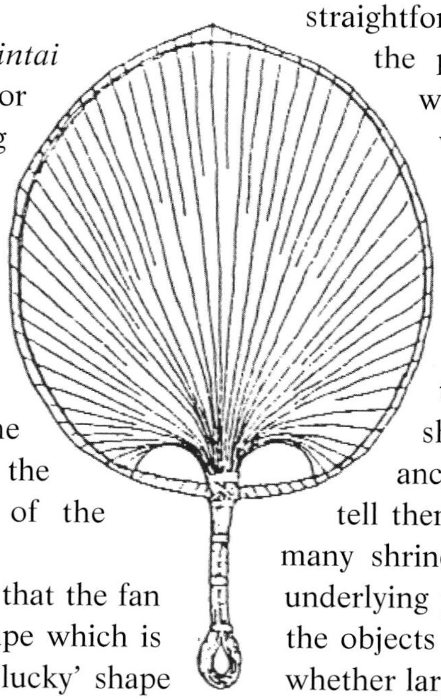

Discussion of the Ancestor–God Imagery and Symbolism

Our ancient ancestors revered the snake as their ancestor–god, as did many peoples around the world, but, unlike the Middle Jomon people, the Yayoi people did not literally depict the snake figure. The objects of their worship were the mountains and forests. The conical shape of the mountain is suggestive of the snake–ancestor in its coiled form, and mountain ridges are reminiscent of the moving, active snake; both were revered. The mountain was also very important as the source of water, and in consequence villages developed in the foothills.

While straight 'phallic' shaped trees and crawling vines were worshipped for their symbolic attributes, in both cases this is not straightforward worship of the god itself, but the practice of 'indirect' worship, in which it is the symbols of the god which are revered. However, this 'symbolisation' is the true essence of Japanese snake worship. With the influence of Buddhism in later periods, shrines came to be erected as the official residences of the *kami*, but even then trees in the shrine enclosure were still ancestor–gods and it was forbidden to tell them. This rule is strictly observed in many shrines to this day. I believe that the underlying principle of the snake cult is that the objects perceived to symbolise the snake, whether large or small, have the same magical powers as the snake–ancestor–god itself.

The snake–ancestor–god was an overwhelming

Fig. 7. An example of a *hoki* fan sold in shops

presence; the legend of the *daidara* (probably big snake) which sits on mountain ridges and looks down on the world; the enshrinement of a single huge *waraji* (woven sandal) probably intended for a snake, and the monstrous *shimenawa* at Izumo shrine and the worship of large trees as sacred, are all examples of the expansion of the snake imagery. Of these symbols, the fan is a secondary image derived from the *hoki* tree; because the sacred trees could not be cut down, the leaf was substituted for the tree, and this is the origin of the fan.

Although there may be other reasons, one of the factors contributing to the abundance of forests surviving in Japan today is the unique practice of *mitate*, or the worship of like objects. Mountains, which symbolised the ancestor–god were forbidden ground, and in forests sacred to the village, it was forbidden to cut down trees. There are snake cults in many areas of the world, and the religious practices specific to the people of each area have left their mark on the surrounding natural environment.

References

Hastings, J. (ed.) (1976): *Encyclopedia of Religion and Ethics*, T & T Clark, Edinburgh, 399 pp.

Yoshino, H. (1986): The Origin of Fan *Ohgi*, Japanese Patterns; Vol. 2, *Fan Ohgi*, ed. S. Imanaga, Shogakukan, Tokyo, 163 pp.

CHAPTER 7

XUTS: Chief of the Woods and the Tlingit of North America

ANNE-MARIE VICTOR-HOWE

I discuss here the unique relationship that exists between the Tlingit people, and the brown bear and the grizzly bear. The role the bear plays in Tlingit stories and narratives will be described and I will examine several Tlingit totem poles and other carvings. The Tlingit culture as it was during the latter half of the eighteenth and early nineteenth centuries, a period when the Tlingit acquired unprecedented wealth as a result of their commerce with European fur traders, will be focused upon.

Migratory people settled along the coast of south-eastern Alaska about 10,000 years ago. According to accounts of Tlingit elders, the Tlingit's ancestral home lay near the mouths of the Nass and Skeena rivers, in what is now western British Columbia. The Tlingit subsequently lost some of their southernmost lands about three hundred years ago, when small groups of Haida from the Queen Charlotte Islands migrated across Dixon Entrance and pushed northward into Tongass Tlingit territory and the southern part of Prince of Wales Island and Dall Island. In the late eighteenth and early nineteenth centuries the Tlingit expanded north across the Gulf of Alaska to Icy Bay and Controller Bay, taking land from the Athapaskans and Eyak in what is now south central Alaska. Since that time, the Tlingit have predominantly occupied the Pacific north-west coast from Puget Sound to Yakutat Bay, in south central Alaska. They are four major groups, from north to south: the Gulf Coast Tlingit, the Northern Tlingit, the Southern Tlingit, and the Inland Tlingit. For the purposes of this discussion, I will primarily be considering the first three, largely coastal Tlingit groups, not the inland Tlingit, whose culture is closer to the Athapaskan way of life.

The coastal environment of the Tlingit people is broken and interlaced with countless fjords, bays, and islands, and is characterised by the relatively cool, wet climate typical of the Pacific north-west region of North America. A combination of high annual precipitation and acidic soils produces a rich, thick vegetation dominated by dense stands of conifers that include spruce, hemlock, and cedar.

In old-growth forest undisturbed by man, the forest floor is carpeted with moss and an understory of shrubs and ferns broken by more open, boggy muskegs. The fauna of the area is plentiful. While marine mammals and a wide variety of salt and freshwater fish were the most important source of food for the Tlingit people during the period of their history we are examining here, land game was also abundant and widespread. Amongst the more imposing species were the grizzly bear and the brown bear. The Tlingit people called both types *Xuts*.

The brown bear (*Ursus arctos middendorfi*) and the grizzly bear (*Ursus arctos horribilis*) are closely related subspecies, and in some areas along the coast of north-western North America their ranges overlap. The Alaska coastal brown bear is found mostly on islands within its range, from Baranof Island in the south to Kodiak Island in north-western Alaska, while the grizzly bear's principal range is farther inland. The brown bear is significantly larger than the grizzly, reaching twelve feet in length and weighing up to 1,500 pounds. The muzzle and face of both brown and grizzly are highly expressive, as are their body movements and vocalisations. They are extremely quick and powerful, able to sprint faster than a racehorse out of a starting gate and to drag an eight-hundred pound moose some distance in their jaws. Their keen sense of smell enables them to detect the presence of humans a mile upwind, and their hearing is so acute they can detect human conversation up to three hundred metres away. Their vision is not as sharp as their other senses, although they can see well at night and in the low light of the temperate rain forest. While they tend to steer clear of people, seldom attacking unless provoked, they are often unpredictable and can be quite dangerous.

Individual brown and grizzly bears have a home range, although they do not attempt to exclude other bears from it; often a male bear's territory overlaps with several of his neighbours. They mark their territories by clawing and biting trees and leaving excretory messages. Adult bears are solitary most of the year, only coming together to court and mate in June and early July. In the case of the brown bear, mating also occurs later in the summer when they gather at streams where salmon, one of their main foods, are spawning.

Brown and grizzly bears are omnivorous, ranging widely in search of a variety of foods, from roots, berries, and grasses to carrion and fish. In late autumn, brown bears dig winter dens in remote, sheltered locations, most often in old-growth forest at elevations of between 1,000 and 2,000 feet, frequently at the base of a stump or wind-felled tree. Some dig dens above the tree line, at higher elevations. For both the brown and grizzly bear, the period of dormancy usually lasts for five to six months. During that time they do not eat, drink, urinate, or defecate, although their sleep is not so deep that they can't become active quickly if disturbed. Pregnant females give birth to their young while in the den, usually between January and March.

Brown and grizzly bears were thus a very real presence in the lives of the Tlingit people for most of the year. The Tlingits admired and respected them as much as, if not more than, any other animal, calling them 'Chief of the Woods' (Emmons, 1991: 133). This was due, in part, to their physical size, strength, and speed, but also to their close resemblance to humans. They often stand upright on their hind legs, like humans, and use their front paws with great dexterity. Also like humans, they eat a wide variety of foods, and often the same high-quality foods most prized by the Tlingit people. Female bears are also devoted mothers during the several years the cubs stay with them before striking out on their own.

In addition to physical and behavioural similarities, the Tlingit saw in the bear's long months of dormancy evidence of its knowledge of the non-human spiritual world. Because it disappeared into the earth in the autumn and re-emerged in the spring, the bear was for them a powerful symbol of death and rebirth. Its isolation and fasting in a dark, remote den bore many similarities to their own initiations, including puberty rites and shamans' initiations.

The Tlingit also believed that bears could understand human language; thus one had to be very careful what one said about them, in case they heard and were offended. The Tlingit hunter was careful to use circumlocutions and metaphorical terms, or kinship terms and honorific names, when addressing them or otherwise referring to them. He had to be both physically and spiritually prepared to go hunting, and was required to fast and remain continent before setting out. He usually ate lightly while hunting, not only for practical reasons, given the distance he often had to travel, but because such frugality and abstinence were believed to be moral virtues with religious or magical significance (de Laguna, 1972: 362). The hunter protected himself with special amulets and medicines. Because women were believed to have special powers over bears, the hunter's wife had to remain quietly at home and not show anger towards anyone, in order to show the bear proper respect, and to help ensure a successful hunt for her husband (de Laguna, 1972: 365).

If the hunter was successful, other rituals and observances were required to show respect for the bear's death and propitiate its spirit. Emmons (1991: 132) noted that after a bear was killed, the hunter sang a special song to it, and its skin was painted with red ochre. Emmons (1991: 133) also recorded an occasion in Sitka, Alaska, in 1894 when Tlingit hunters killed two brown bears and, after stretching their skins out to dry, covered the bears' heads with eagle down to honour their spirits. Sometimes the bear's skull and the bones of its feet were either buried in the ground or thrown into the sea. De Laguna (1972: 365–6) describes another, somewhat similar ritual performed by Tlingit hunters in Yakutat. These hunters cut off the head of the bear and buried it facing east, toward the rising sun.

Some of the most revealing insights into the Tlingit people's relationship with brown and grizzly bears come from several widespread myths and stories. *The Woman Who Married the Bear* is a well-known story in western Canada and Alaska. The following is my summary of a version told by Louis and Florence Shotridge (Barbeau, 1990: 211–3):

> There once lived a chief who had many sons and one beautiful daughter. Many young men wanted to marry her, but she refused for one reason or another. One day she went berry-picking in the woods with her friends. On their way home, the chief's daughter stepped into a bear's track and slipped. She then made some uncomplimentary remarks about bears: a mistake, since the spirit of the bear could hear her and be offended. They all started walking again, but when the strap of the girl's basket broke, she told the other girls to continue without her, and that she would catch up with them later.
>
> It was already dusk when a handsome young man offered to help the girl. When she accepted his offer, he picked up the basket and asked her to follow him. Late in the evening they reached a village, but it was not the girl's home. The young man then took her to his father's house in the centre of the village, where he introduced her to his father and told him that the girl was his wife. The father, a person of considerable rank in the village, gave a feast in their honour.
>
> After a while, the woman started noticing strange things about her husband and the other people in the village. When the men returned from hunting and shook their wet coats in front of the fire, she observed that the fire burned more brightly. She finally asked her husband to let her go with him the next time he went to his fishing camp. In camp she observed that while she was gathering dry firewood, the other women only picked up water-soaked wood. She was humiliated later when her husband tried to dry himself by the fire she had built with her dry wood, and the drops of water from his wet coat extinguished it. She was even more ashamed when she noticed that the other wives' fires remained burning while their husbands dried their coats. She knew then that something mysterious was happening.
>
> One night after they had returned to their village, the girl had trouble sleeping, and in the middle of the night discovered a large grizzly bear lying in bed next to her. She realised then that she was among bear people and that she had been taken away from human society because she had spoken disrespectfully about the bears after slipping on their tracks in the woods. She wanted to run away, but she could not escape.
>
> Meanwhile, everyone in the Tlingit village thought she was dead. In early spring, her brothers went hunting and soon came upon the place where their sister was living with her bear-husband. The girl heard her brothers outside the bear den, so she called her bear-husband and told him not to fight because the men outside were his brothers-in-law. In the ensuing fight, the bear-husband was killed, giving up his life because he knew he had been wrong to take the girl. When she went out she saw the bear lying with arrows in its side and her brothers about

to begin skinning it. She told them to stop because he was their brother-in-law. Her brothers were surprised to see her. The girl told them what had happened to her. The men buried the bear and took their sister home, leaving behind her two sons, the cubs with half-human faces.

This story of *The Woman Who Married the Bear* has much to tell us about how the Tlingit people perceived their environment and the dynamic relationship between the secular and sacred worlds. For the Tlingit, the secular, earth domain was inhabited by humans and their tribal ancestors, as well as certain animals, including bears, which also lived in the sacred world. Beneath the earth, and in the sea and sky, were sacred realms ruled by non-human, supernatural forces.

The ocean and the forest were both part of the earth realm, although the ocean was more familiar to the Tlingit. They were a coastal people who built their villages along the shore and relied on the sea for much of their livelihood. The forest was less important to them economically, although it provided them with game, animal skins they could trade with other parties, and wood to build houses, canoes, and many different objects used in everyday life. The Tlingit also used a wide variety of forest plants to prevent and cure sickness.

Yet the forest was perceived by the Tlingit to be a strange, often dangerous place inhabited by supernatural beings who, like the bears in *The Woman Who Married the Bear*, disguised themselves as humans. In the forest it was dark, the lush vegetation was overwhelming, and one could not see very far ahead. With the mist and fog, and the shifting, patchy light, the environment was often only half-revealed and constantly changing. Human perception was distorted. The song of a bird might come from a bird, but one could not see the bird, so the song might come from something else. Natural objects could suddenly look like something entirely different. Like the girl in the story who thinks that she has just met the most handsome man in the world, it was hard to spot the disguises of supernatural beings in the forest. It was only after living in her bear-husband's village deep in the woods for a while, far from her village and other human beings, that she began to notice strange things happening around her: the women gathering wet wood for their fires, for example, and water from their husbands' wet coats making fires burn brighter. Thus the story also shows how in the supernatural world of the Tlingit, natural phenomena were often reversed.

Another version of *The Woman Who Married the Bear* (Dauenhauer and Dauenhauer, 1987: 177) illustrates the bear's spiritual powers. In that version, the bear-husband is a shaman. When the girl begins to miss human company and to think of escaping, her husband's special powers allow him to discover his wife's plan. He also foresees his death at the hands of her brothers.

The Woman Who Married the Bear also addresses certain Tlingit conventions having to do with marriage. While Tlingit culture required marriage outside the clan, marriage to a bear was far too distant to be workable and was thus doomed to end tragically. When the woman discovers that she is living with bear people, she wants to run away, but cannot because after three years with her husband she has come to love him and she is devoted to their children. In a version told by Tlingit elder Tom Peters and recorded by Nora Dauenhauer in 1972 and 1973 (Dauenhauer and Dauenhauer, 1987: 189, 191), as in other Tlingit stories about people who live for a while in the supernatural world, the woman finds that she cannot go back to her own kind. When she wears her bearskin she turns into a bear and takes on all the features and powers of that animal.

Yet, while *The Woman Who Married the Bear* has much to do with the strangeness and dangers of the supernatural world, it also highlights the Tlingit view that the animal world is not altogether different from the human world. As

Jonaitis (1986: 96) points out, to the Tlingit the bear is an important symbol of the human social order. The bear people live in a village much like a Tlingit village, they go fishing and hunting, and when spring comes they prepare to leave the winter village for their summer camps. Like the house of a Tlingit chief, the bear-husband's house is in the middle of the village and thus signifies his high rank. Like a Tlingit chief, the bear-husband's father organises a big feast in honour of his son and new daughter-in-law. In choosing to sacrifice himself for the benefit of the clan, and in a version of the story in which he instructs his wife in the proper observances and customs after his death, the bear-husband seems guided by human rules (McClellan, 1970: 8; Dauenhauer and Dauenhauer, 1987: 185). Indeed, as McClellan (1970: 7) points out, 'the bear-husband comes out as a truly noble character, for he properly fulfills his role as brother-in-law.'

Fig. 1 shows a Tlingit totem pole at Wrangell, in south-eastern Alaska. While it is not directly related to the story of *The Woman Who Married the Bear*, it clearly indicates that the Tlingit saw bears belonging more to the social order than to the sacred realm. The bear is standing upright, like a human, and wears a hat crowned by a stack of rings. In Tlingit society, such a hat signified great wealth and status, and each ring represented a potlatch or ceremonial feast given by the master of the house who was the custodian of the hat. By giving frequent and elaborate potlatches, and by publicly displaying the clan's crests, the wealth and status of the master of the house (or *yitsati*) and his clan were maintained and enhanced. The presence of such basketry rings atop the bear on this Tlingit totem pole thus signifies that the bear belonged more to the human social domain than to the spiritual realm: the master of the house is chief of his house, and the bear *Xuts*, is 'chief of the woods'.

The relationship between Tlingit shamans and the bear is also revealing, as the shaman was the member of the Tlingit community who could cross back and forth most successfully between the natural and supernatural worlds. One ivory charm illustrated in Emmons (1991: 367) represents a shaman's dream of a bear biting one man and holding another. Probably it was meant to symbolise the shaman's rite of passage into the sacred realm, when he underwent death and rebirth. Again the human is inverted and, as with

Fig. 1: Totem Pole: The Bear at Wrangell, Alaska. Neg. No. 46127 (Courtesy Department Library Services, American Museum of Natural History, New York)

the house post collected on the Harriman Expedition in 1899 at Cape Fox, Alaska, the figure's position might represent the rite of passage of being eaten and dying and being reborn. The inverted position could also represent the reverse world of liminality.

Although the majority were male, Tlingit shamans could be of either sex. Shamans were very powerful, and had many important responsibilities in Tlingit society, including curing the sick, assisting in warfare, assuring the welfare of people by protecting them against witches and other malevolent spirits, foretelling the future, and assuring an abundance of food by assisting the hunters.

The shaman acquired his (or her) power during several weeks of training and initiation in a remote location far from the village, with only a small amount of food. Kamenskii (1985: 83) gives a particularly lyrical, subjective description of a shaman initiate's experience during a vision quest:

> Face to face with the wild, mysterious, and formidable nature, [the shaman] begins his exploits. Century-old spruce trees and cedars that prevent sunlight from penetrating the forest serve as his cover. When everything is quiet, and only the sound of the surf lightly touching the rocky shore can be heard, these ancient giants slowly sink into a lazy slumber. Only somewhere up above, their majestic crowns whisper lightly, rustling their branches. But how terrifying is their cracking noise when the gusty wind starts hitting their heads. This mysterious and frightening noise, along with the rumble and roar of the ocean waves crashing against the rocks, makes every living creature tremble. Animals and birds hide in their shelters. A shy deer seeks refuge among the rocks, under the cover of high cliffs. Every creature is terrified, but the future shaman does not share this feeling. He listens carefully to the howling of the storm and the noise of the forest and the sea, as to voices from that mysterious world that he intends to apprehend. This is the voice of the elements that he wants to learn to dominate and command.

According to Tlingit oral histories, after a week or more of fasting, an animal would usually cross the shaman initiate's path, extend its tongue, and fall dead. The tongue was considered to be the conduit for the animal's knowledge and the conduit through which life passed from one being to another, so by presenting its tongue it was agreeing to pass its knowledge and life on to the novice. By cutting out the tongue and keeping it, the novice acquired the animal's knowledge and spirit: its *yéik*. (Jonaitis 1986: 52–53).

The land otter was always the first animal to appear to a novice shaman. It was perceived to be a dangerous, supernatural beast because it could transform human beings into its own kind. It also had magical qualities that could assist the shaman on his journey. As suggested by Jonaitis (1986: 92) 'the shaman had two options for handling it: he can try to conquer it, as he does the witch; or he can try to coerce it to work for him, rather than against him'. Often he did so by cutting out the otter's tongue, tying it up in a bundle made of two pieces of wood and hiding it away in the woods in a cave or in a hollow tree (de Laguna, 1972: 677). It was then safe for him to encounter other animals, including the bear. Some particularly powerful Tlingit shamans had as many as eight such hidden tongues (de Laguna, 1972: 677–78).

As a spirit helper for the shaman, bears were perceived to be quite different from land otters. Unlike the land otter and several other animals that embodied only the spirit world, bears (along with wolves, ravens, and several other animals) often appeared in both a sacred and secular context within Tlingit culture. Bears could travel back and forth easily between the secular and the sacred realm and were equally at home in each. Thus they were welcome helpers to the shaman both during his quest and upon his return from it, when he had to fit back into human society.

Like a bear emerging from its den in the springtime, by the time the shaman returned to his village he had lost considerable weight and was very weak. The initiation of the shaman, or the vision quest in the woods, was thus seen to be a kind of death and rebirth much like the winter dormancy and springtime emergence of the bear. Thenceforth, during shamanic ceremonies, the shaman was able to transform himself into his spirit helper by wearing different masks that represented each of his *yéik*s, including that of the bear if he had acquired its powers.

Largely because it could pass so easily back and forth between the secular and sacred worlds, the bear was one of the principal crests of several major Tlingit clans of the Wolf (also Eagle) moiety. Crest were sets of emblems that traced the connection of an individual and his group to the mythical world. They were among a Tlingit clan's most valued possessions. They often changed hands through marriage and potlatches, and were passed on to succeeding generations. It was believed that they had been acquired as the result of real events or in the dreams and visions of ancestors who had encountered certain beings in animal form, and had received supernatural powers from them. Thus these beings became *at.óow*, part of the real or incorporated property of a particular clan, and positioned an individual in a moiety and lineage, and symbolised its owner's social identity. A clan's *at.óow* might also include songs, stories, artistic designs, personal names, land, and other elements (Dauenhauer and Dauenhauer, 1990: 13). Tlingit crests were carved, painted, and woven into totem poles, house screens, house posts, masks, wooden hats, dance head-dresses, helmets, daggers, ceremonial wooden bowls, Chilkat blankets, button blankets, and other costumes. Some smaller, portable crest objects were displayed or worn in ceremonies, including memorial ceremonies held to pay opposite moieties for services rendered during and after a funeral (Jonaitis, 1986: 42).

Before firearms were introduced by American and European fur traders in the nineteenth century, the dagger was the most important weapon of a Tlingit warrior. The ivory handles of daggers often represented the head of a grizzly bear. By carrying such an object with a bear crest on it, the warrior had the strength and supernatural powers of the bear—and in the woods no other animal could challenge the bear; it was invincible.

A wooden screen illustrated in *The Tlingit Indians* (Emmons, 1991: 32) shows a bear crest on the front of a Ground Shark House at Wrangell. It shows all phases of the rite of passage. The entrance to the house is in the bear's belly. By entering the bear's belly a person gained its protection while inside the house. Inside many Tlingit houses such as this, there were often similar screens representing the crest animal of the house. Only higher-ranking individuals, most notably the master of the house or *yitsati*, could go behind it, again entering through a hole cut in the screen-being's belly. According to Jonaitis (1986: 133):

> The hole in the stomach of a crest animal on an interior screen conveys all three phases of a rite of passage, probably because its very use as a screen through which an individual passes constitutes a ritual. Placed on the belly on the animal, the hole in the stomach of the screen-being is like a passage into its womb; when the chief enters into that womb, he becomes engulfed by the crest being, and separate from the profane world. While behind the screen, the chief is in the belly of the animal, enjoying the liminality [the threshold of consciousness] of temporary unification with the crest animal. When he returns to the central area of the house, he does so by coming out of the mythical animal's womb, and is thus incorporated back into his social group by the metaphor of birth .... The hole in the stomach of a screen permitted a *yitsati* to undergo a three-phased rite of passage that revalidated his social position.

Another Tlingit house at Gash village has a bear crest ornamenting its entire front. The bear has an inverted head of a man in its jaws. The man's face is also portrayed above the bear's nostrils, between its eyes. The two positions of the man suggest that he was not an ordinary man, but could enter the reversed world of the supernatural. The two human heads might also signify that bears could supernaturally change and take human form, and similarly that humans could transform themselves into bears. Here that transformation is indicated by specific designs on the bear. The lines on the bear's body suggest the ribs of the animal, and might also suggest death and rebirth, as the Tlingit people always associated bones with death and rebirth—in this case, the regeneration or rebirth of the bear.

Some crest items were only shown or worn on special occasions. Two such items, a Naaya.aayi grizzly bear mask and a costume worn by a performer, appear in a photograph taken by Niblack in 1885 (Emmons, 1991: 174). That mask and bear skin costume were among the most precious *at.óow* of the Naaya.aayi clan. On the mask (now in the collections of the Burke Museum, Seattle, Washington), the raw skin of a bear has been stretched over a wooden armature. The eyes of the bear are made of iron, the lips are of copper, and the teeth are of opercula and bears' canine teeth. The ears have been emphasised with humanoid bear's faces worked in repoussé on fine copper sheets. The eyes and teeth have been inlaid with abalone (Holm, 1987: 192). The mask is a good example of how the lips, eyes, eyebrows, or ears of Tlingit bear masks were often elaborately decorated with copper and other precious material in order to accentuate those parts of the animal and associate both the bear and the wearer of the mask with wealth and prestige.

Another photograph shows Chief Shakes V, who was from the Stikine River area, lying in state, wearing or surrounded by all the most important *at.óow* of the Naaya.aayi clan (Emmons, 1991: 272). The bear mask and costume described above are displayed in the picture as a bear standing up in a lifelike position. This bear costume was also worn by clan members during important potlatches and ceremonies. One of these ceremonies enacted the escape of Chief Shakes' ancestors from a great flood in the company of a bear and a mountain goat. Since that time there has been an alliance between the bear and the Naaya.aayi clan of the Stikine wolf phratry. The following is my own summary of a version of that story recorded by Swanton (1909: 231):

> At the time of the flood the Naaya.aayi were climbing a mountain along with a grizzly bear and a mountain goat. Finally they killed the bear and preserved its skin, claws, teeth, and so forth, intact. They kept it for years and each time when it began to deteriorate they killed another bear. This is how they claimed the grizzly bear crest. When the bearskin was shown thousands of dollars worth of slaves and furs were given away. They were very proud of owning this bear and did all kinds of things toward it. Many songs were composed about it. One of them said 'Come here, you bear, the highest bear of all bears'.

The Teikweidí and the Kaagwaantaan clans also claim the bear as a crest animal. The Teikweidí were said to have acquired their name at a peace ceremony given for the Dog Salmon:

> Bears killed all the Dog Salmon until there was only one salmon left. That salmon swam up a river but soon was captured by bears who then took him to their camp. The bears then invited all the Teikweidí, and all the people living at Mountain-Inside-Town to a peace ceremony. During the ceremony they painted the salmon with the red stripes which all Dog Salmon still wear to this day. The bears also made death payment for all the relatives of the salmon they had killed, and agreed not to kill so many salmon in the future so that the tribe would remain prosperous. During the ceremony the bears also instructed all hunters present in the proper treatment of bears. They told them that they should decorate the head of the bear with red stripes and eagle down, and taught them a song they should sing to it. [Garfield, 1947: 443–4].

A Tlingit chief's wooden staff with a bear carved

into it illustrates the importance of the bear as a peacemaker (Fig. 2). Such staffs were used to command warriors and to settle arguments and separate parties engaged in disputes (Jonaitis, 1986: 23). The staff gave the chief the power to do so, and was highly respected. It was brought out on very special occasions and was only carried by high-ranking individuals in Tlingit society.

Another story describes the origin of the bear crest of the Kaagwaantaan:

> People used to catch herrings and harvest herring eggs in Neva Strait, near Sitka. One night a big bear stole all the drying eggs of an old widow. The next day the widow hung more eggs out to dry. She only saw one big hand reaching for the eggs. The eggs were stolen again and the upset woman called the one who stole her food a thief. The bear then killed her and several other people. The men in the village tried to stab him but could not succeed so they went to Sitka and told their story. All the Kaagwaantaan decided to hunt the bear. One young man stabbed the bear from behind but as he was doing so he fell over backwards. As the bear prepared to jump on him he raised his spear and when the bear landed it impaled himself. The other man came to rescue the young man still on the ground and stabbed the bear. They skinned him, cut off his ears, and took his teeth and claws [Olson, 1967, p. 40].

In both these stories of the Teikweidí and Kaagwaantaan, as in the story of *The Woman Who Married the Bear*, the bears act nobly, according to the rules of proper human behaviour. In the first story, the bears are sympathetic peacemakers who help the Teikweidí and ask in turn that the clan perform the proper rituals and ceremonies to show their respect for bears. In the second story, although there is violence between people and bears, in the end the bear, in spite of his supernatural powers, voluntarily sacrifices himself by falling on the Kaagwaantaan hunter's spear, thus allowing the clan to acquire a bear crest.

The story of Kaats', a Tlingit man killed by a

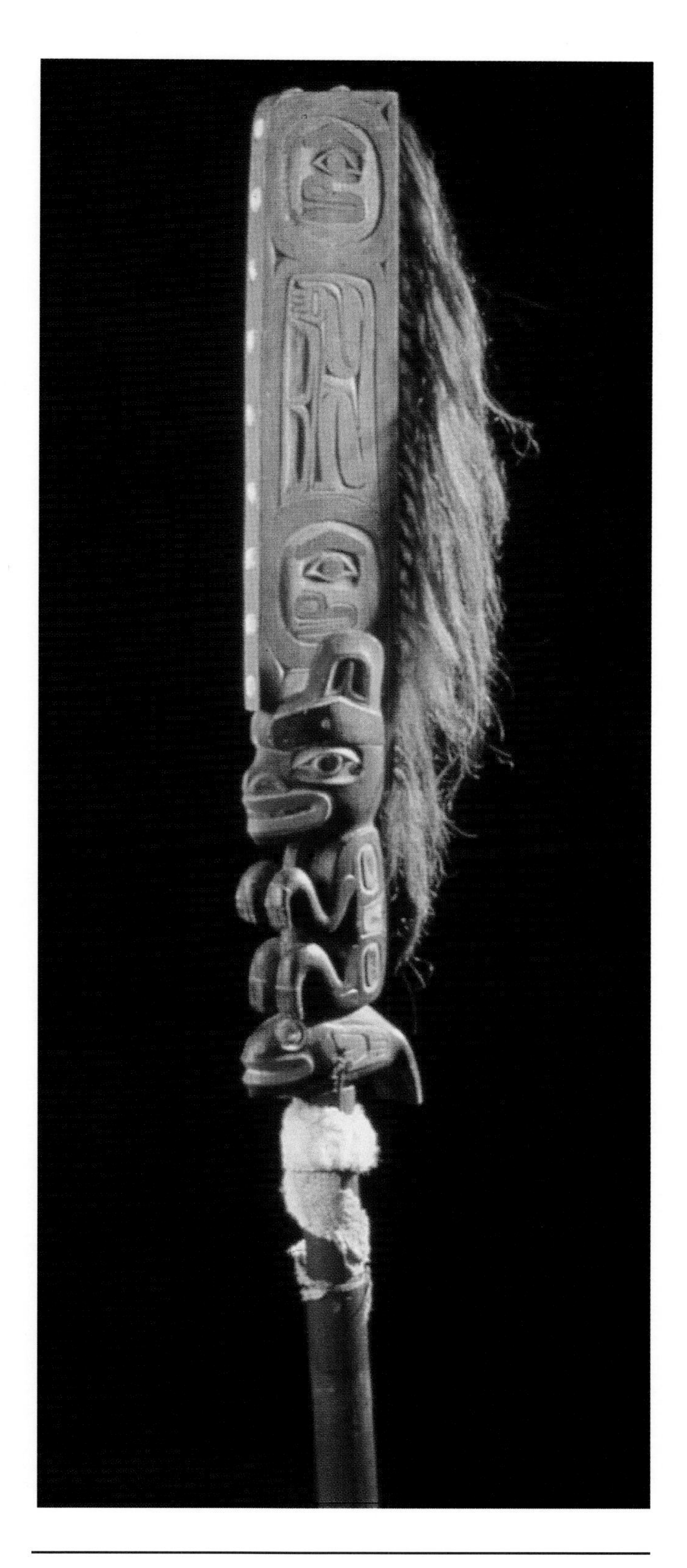

Fig. 2: Chief Staff with Bear and Killer Whale. Photo by Hillel Burger–S10906 (Courtesy Peabody Museum of Archaeology and Ethnology, Harvard University, Cambridge)

bear, also describes a violent encounter between people and bears in which the bears act at least as nobly as people. Scholars have recorded a number of different versions of the Kaats' story, all with slightly differing story elements, plots, and characters (Swanton, 1909: 49, 228–9; Garfield and Forrest, 1961: 29–37; Keithahn, 1963: 156; Barbeau, 1964: 215; de Laguna, 1972: 879–80; Dauenhauer and Dauenhauer, 1987: 219–43). In a version recorded by Swanton (1909: 49), Kaats' belongs to the Kaagwaantaan. Swanton was of the opinion that the version he recorded described the Kaagwaantaan's acquisition of its bear crest. The following is my summary of that version:

> While out hunting with his dogs, Kaats' was captured by a she-bear. She pulled him into her den, concealed him, and then married him. They had several children. Inside the den the she-bear and other bears took off their skin coats and looked just like people. After a while Kaats' began to miss his own kind and wanted to return to them. The she-bear agreed to let him go, but on one condition: he must not smile or touch his human-wife or hold his human-children in his arms.
>
> Once back home, Kaats' hunted seals and other marine mammals, giving some of his catch to his bear-wife and children. One day, however, he held one of his human-children. His bear-children saw him and killed him. The bear-children then scattered and hid far and wide, but all except one were eventually killed by Tlingit hunters. The surviving bear-child then heard a girl say something in anger about him, so he killed her and everyone else nearby except for a few people who escaped in their canoes, and one woman whom he captured and carried off. Some men finally managed to kill the last of Kaats' bear-children with spears and knives.

Fig. 3: House Post representing a bear holding an inverted human in its jaws. Collected on the Harriman Expedition in 1899. 333 cm x 92 cm x 31 cm. Cat. No. 1-10831 (Courtesy Thomas Burke Memorial Washington State Museum, Seattle)

The story was memorialised in a totem pole now on display in Saxman Park, near Ketchikan, in south-eastern Alaska. The blending of human and bear imagery on that pole illustrates the central point of the story: the bear's easy movement between the sacred and secular realms. The main figure is Kaats', carved in large relief. Nestled on his head, between his ears, is his much smaller bear-wife. Garfield and Forrest (1961: 30) point out that while Kaats's animal ears make him look less human, they also indicate that he has supernatural or divine powers. He is holding one of his children in his hands, and the child appears to be human. The position of the bear's wife, who is sitting with her paws resting on Kaats's ears, is also very human.

Fig. 3 shows a carved and painted post from a Tlingit house. It was collected by the Harriman Expedition in 1899 at the abandoned Tlingit village of Gash, at Cape Fox, Alaska, and may have come from a house there called the 'Kaats' House.' It shows a human, presumably Kaats', issuing head-first from the mouth of a bear, as if he were either being eaten or born, or both. Thus it could represent Kaats's rebirth in the spirit world of the bear, or his sexual union with his wife, as in Tlingit culture the act of eating is often associated with sexual exchange. The figure's inverted position might also indicate the reverse world of liminality as described above in my discussion of *The Woman Who Married the Bear*.

It is worth noting that like *The Woman Who Married the Bear*, Kaats' is captured by a bear while he is in the forest, but he does not realise that his wife is a bear until they are deep in her den in the woods, far removed from human society. A Kaats' story recorded by Nora Dauenhauer is a different version of how the Teikweidí clan acquired its bear crest. As in *The Woman Who Married the Bear* and Swanton's version of the Kaats' story, the crest is acquired at the cost of human life. Most noteworthy about the Dauenhauer version is that it strongly emphasises the role of supernatural illusion—even the bears' den looks like a Tlingit house to Kaats' for a while after he arrives (Dauenhauer and Dauenhauer, 1987: 221). Dauenhauer's informant, J. B. Fawcett, also included in his narrative more instruction on the proper, respectful way for humans to behave towards bears, and described in greater detail bears' human-like emotions: the deep affection the bear-wife has for Kaats', for example. 'She was kind to him,' Fawcett, a Tlingit elder, told Dauenhauer. 'If only things hadn't happened this way, how would it have been? It would have really been something, they say. That's how it's told, you see. This is why the brown bears understand humans' (Dauenhauer and Dauenhauer, 1987: 237). As Dauenhauer comments, 'This passage is one of many in the story wherein the narrator meditates on the relationship between humans and bears, and how much the bears understand us, have compassion, and are like us, but that humans historically have been unable to reciprocate. The relationship is now delicate and often ambiguous' (Dauenhauer and Dauenhauer, 1987: 403).

Whether a bear hunter or a practicing shaman, one had to negotiate with 'The Chief of the Woods' and treat him with respect if one wanted to hunt successfully, grow prosperous, or travel safely between the sacred and secular realms. In the last analysis, the bear was a foil or mirror for much of what the Tlingit people understood or sought to understand about their secular and sacred worlds. The bear was even present on their journeys into death which, in Tlingit cosmology, involved a trail through the woods that led from the village cemetery up into the mountains. The bear, *Xuts*, was there beside the trail, every step of the way: a stranger and a friend, a neighbour and a distant acquaintance, a foe and a helper, a warrior and a peacemaker, a source of both understanding and confusion, an imposing guide on the final journey into the sacred realm.

References

Barbeau, M. (1990): *Totem Poles According to Crests and Topics*, Canadian Museum of Civilization, Quebec, Canada.

Dauenhauer, N. M. and R. (eds.) (1987): *Haa Shuka, Our Ancestors: Tlingit Oral Narratives*, Classics of Tlingit Oral Literature 1, University of Washington Press, Seattle.

—— (eds.). (1990): *Haa Tuwunáagu Yls, for Healing Our Spirits*,

Tlingit Oratory, Classics of Tlingit Oral Literature 2, University of Washington Press, Seattle.

de Laguna, F. (1972): *Under Mount Saint Elias: The History and Culture of the Yakutat Tlingit*, Smithsonian Contribution to Anthropology 7, Washington: Smithsonian.

Emmons, G. T. (1991): *The Tlingit Indians*, ed. with additions by F. de Laguna, University of Washington Press, Seattle.

Garfield, V. (1947): Historical Aspects of Tlingit Clans in Angoon, Alaska, *American Anthropologist,* 49 (3): 439–52.

Garfield, V. E. and L. A. Forrest (1961): *The Wolf and the Raven*, University of Washington Press, Seattle.

Holm, B. (1987): *Spirit and Ancestor: A Century of Northwest Coast Indian Art at the Burke Museum*, Thomas Burke Memorial Washington State Museum Monograph 4, University of Washington Press, Seattle.

Jonaitis, A. (1986): *Art of the Northern Tlingit*, University of Washington Press, Seattle.

Kamenskii, Fr. A. (1985): *Tlingit Indians of Alaska*, trans. Sergei Kan, The Rasmussen Library Historical Translation Series 2, University of Alaska Press, Fairbanks.

Keithahn, E. (1963): *Monuments in Cedar*, Superior Publishing Company, Seattle.

McClellan, C. (1970): *The Girl Who Married the Bear*, National Museum of Canada Publications in Ethnology 2, Ottawa.

Olson, R. L. (1967): *The Social Structure and Social Life of the Tlingit in Alaska*, Anthropological Record 26, University of California Press, Berkeley.

Swanton, J. R. (1909). *Tlingit Myths and Texts*, Bureau of American Ethnology, Bulletin 39.

■ CHAPTER 8

Eyes of the Forest Gods

YOSHINORI YASUDA

Medusa, Who Became a Monster

■ A TALE OF ERADICATING MONSTERS

Medusa is a monster from Greek myth; her hair was made up of snakes and she had a terrifying gaze. People who looked into her eyes would be so petrified that they instantly turned to stone.

It was the Greek goddess Athena who turned Medusa into such a loathsome figure: until then she had been a beautiful virgin. Medusa's mother was Ceto (a sea monster), daughter of Gaea (the Earth Mother). Ceto married her brother Phorcys (the old man of the sea), and bore strange monsters: the three old women, Graeae, and the three Gorgon sisters of whom Medusa was the youngest.

Having heard of Medusa's beauty, Athena decided to compete with her in terms of who was the more beautiful. Athena, however, lost the competition, and in revenge transformed Medusa into a fearsome monster. Medusa's luck continued to plummet, and next, her life was taken by Perseus, son of Zeus, with help of goddess Athena, who armed him with a shield of burnished bronze and warned him never to look Medusa in the eye, no matter what the provocation.

Donning an invisible cloak given to him by a nymph, Perseus became invisible and, approaching Medusa while viewing her image reflected in the shield he had received from Athena, sliced off her head at a stroke with the golden sickle given to him by Hermes. The blood flowed profusely from Medusa's severed head, and then from it sprang the winged horse, Pegasus. This is the tale of the destruction of Medusa, related in Greek myth.

What had Medusa done to deserve this? She had been transformed into a hideous monster, and lived like a worm wriggling furtively underground; why did she have to be killed? Why was Medusa transformed into a monster just because she was beautiful, and why in the end had she to be killed? To date, nobody has seriously questioned the killing of Medusa in Greek mythology.

This question parallels that of why the living creatures of the earth, and the snake in particular, had to be persecuted and killed by humans. This question has been at the back of my mind for a very long time. Why is it that animals have to be killed so very cruelly by humans?

The objective of this chapter is to answer these two questions and reveal the ugly ramifications of global environmental problems today.

■ MEDUSA'S TERRIFYING VISAGE

Medusa was already known in the eighth century BC, the height of Greek civilisation. The bas-relief of Medusa depicts her as a Centaur. The relief is carved into an amphora (now in the Louvre in Paris) unearthed from the Boeotia region in a

Fig. 1: The bas-relief of Medusa depicts her as a Centaur carved into an amphora (Louvre in Paris. Photo by Yasuda)

suburb of Thebes in Greece (Fig. 1). With his right hand, Perseus is aiming the sword he received from Hermes at her neck, and with his left hand he is grasping Medusa's hair. Perseus is also averting his eyes from Medusa's gaze, her eyes are bulging, and she has large ears. She is naked from the waist up, her breasts bared but her lower half is covered by a wrap resembling a skirt. The amphora is tucked away in a corner of the Louvre and from the vast multitudes who visit the museum few pay any attention to the amphora which lies silent and neglected.

In Greek art history, the majority of Medusas, from those of geometric shapes to the so called Archaique of the early period have a body. Within the Medusa, depicted on the amphora from the ruins at Eleusis to the south-west of Athens, lurks a clownish nature under the fearsome visage (Fig. 2). This round-faced Medusa, whose face resembles a tripod kettle or an upturned earthenware pot, is innocent, with somewhat clownish eyes. This urn, made somewhere between 670 and 660 BC, held the bones of a young boy. However, this Medusa too has long hands, and she wears a skirt. The reason we know that this is Medusa with her elder sisters, the Gorgons, is because snakes can be seen emerging from their round faces. At the end of their bodies,

Fig. 2: The round-faced Medusa painted on an amphora from the ruins at Eleusis, Greece (Eleusis Archaeological Museum. Photo by Yasuda)

some of the snakes have animal faces resembling those of lions.

There is no sense of lively motion in these early Medusas, be it that unearthed at Boeotia or the one depicted on the Eleusis amphora. The Medusa they depict is perfectly still, as if time had stopped momentarily. On the other hand, there is the Medusa in violent action as depicted on the western gables of the temple of Artemis at Corfu in north-west Greece built around 600 BC. This Medusa has a fearsome appearance and in this depiction her eyes bulge, her tongue lolls out, and her canines are bared. As if ready to leap upon you at any moment, she is running with her feet spread wide, and fastened around her waist is a belt of snakes. The height of this Medusa is 3.5 metres, and has been claimed to be the first giant

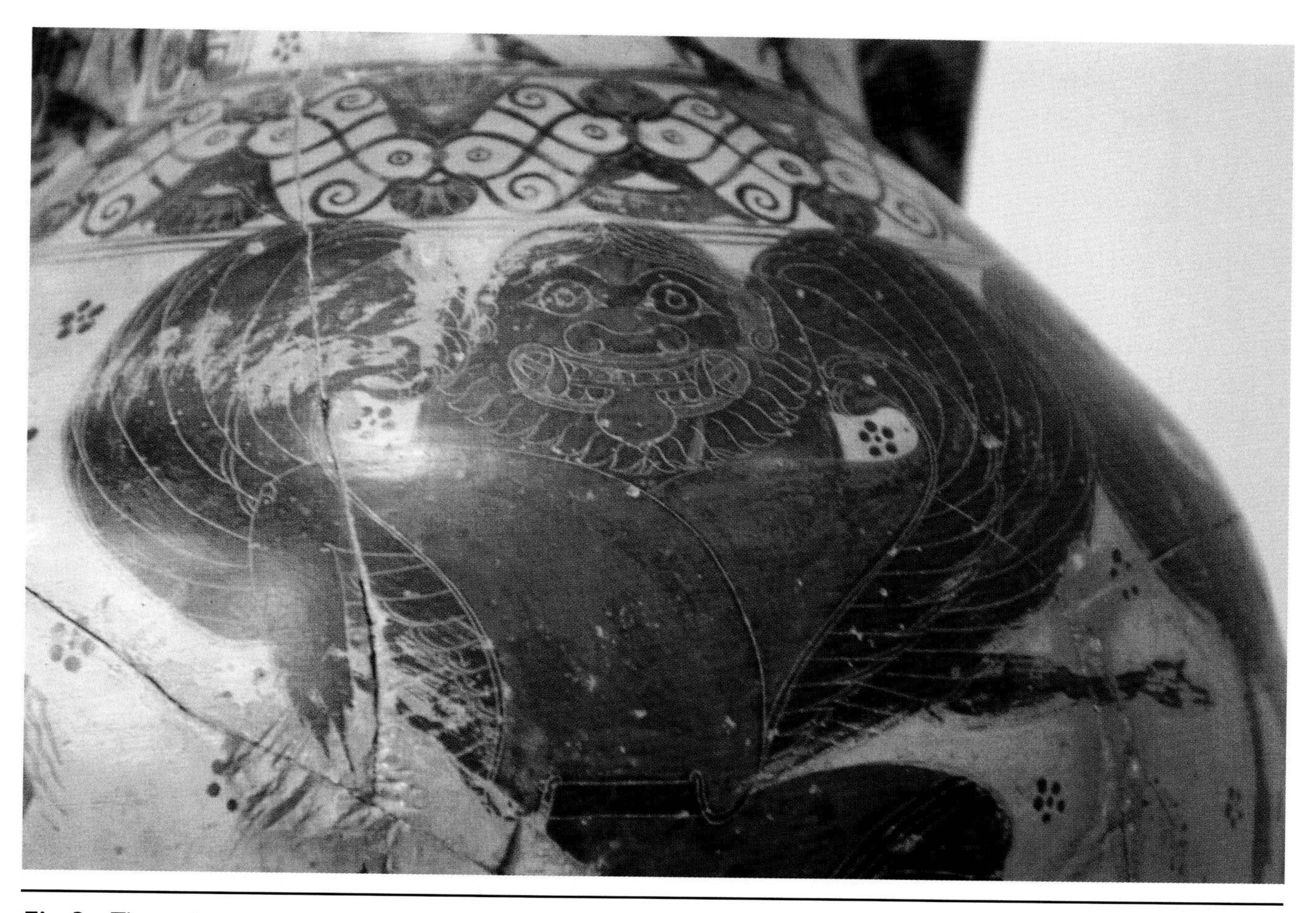

Fig. 3: The violently active Medusa on an amphora found at Nethos, Greece (National Archaeological Museum, Athens. Photo by Yasuda)

bas-relief carving from Greece (Fukube, 1987).

This violently active Medusa is also depicted on an amphora found at Nethos (Fig. 3) which is thought to have been made at around the same time. On display at the Athens National Museum of Archaeology, Medusa is depicted in black on a great urn almost the height of an adult. This Medusa, and her sister Gorgons, have feathers. With huge bared teeth and her hair blown back as she runs, this Medusa is still somehow comical. There is a picture of the feathered Medusa in Turkey too. A work dating to about 530 BC and on display in a museum in Istanbul, a stone Medusa unearthed from the ruins at Didima in Turkey, shows a Medusa with feathers (Fig. 4). The mouth of this Medusa is opened wide and her bared teeth lend her a fearsome appearance.

The period of Greek art history of the Medusa both from the geometrical forms to that of the

Fig. 4: The Medusa with feathers unearthed from the Didima site, Turkey (Istanbul Archaeological Museum. Photo by Yasuda)

Archaique form is depicted with a fearsome appearance, a body, and feathers. In the era when this terrifying Medusa form was created, people felt cowed by the awe of nature, and it was this awe that inspired the fearsome figure of Medusa.

■ BEAUTIFIED MEDUSA

By defeating the Persian army at the Battle of Plataea and the Battle of Mykale in 479 BC, the Greek army swept Persian influence out of the Aegean Sea and ushered in an era of flourishing Greek civilisation, centred on Athens. Along with individual self-reliance, the difference between rich and poor increased substantially. Developments could be seen in mathematics, philosophy, the sciences, and culture, producing in rapid succession Socrates, Herodotus, and Sophocles, among others. In the field of art, realism flourished, beginning the era of classical form.

Since this era, the formerly terrifying Medusa has suddenly become beautified, artists emphasising her feminine charm, and many of these depictions are from the neck up only. These beautiful busts of Medusa were passed on to the eras of Hellenism and Rome. Typical of this are the Medusas from the Didima ruins in Turkey (Fig. 5, Fig. 6) one example of which is smiling gently (Fig. 7). Why then was Medusa suddenly beautified like this? I believe that behind this lies a major change in the relationship between humans and nature.

The cultural developments symbolised by the construction of the Parthenon and the increase in population destroyed the forests. During the Persian wars, large quantities of lumber were felled to construct warships, and then the prosperity of the Greek world that followed added to the destruction of forests and woodlands. The darkness of the forest was opened up, and once vast tracts of land were exposed to the bright rays of the sun, there was room only for human rationality and knowledge and fear of nature gradually receded. Thereafter the idea of

Fig. 5: The Didima Temple in Turkey (Photo by Yasuda)

Fig. 6: The beautiful and graceful Medusa from the Didima Temple, Turkey (Photo by Yasuda)

Fig. 7: The beautiful Medusa carved on the stone coffin (Antalia Archaeological Museum, Turkey. Photo by Yasuda)

Fig. 8: The beautiful Medusa carved on the stone coffin (Konya Archaeological Museum, Turkey. Photo by Yasuda)

anthropocentrism dawned in people's minds, and liberated from the fear of nature, artists began depicting Medusa as beautiful (Fig. 8).

■ UPSIDE-DOWN MEDUSA

Reappraising the notion of a fear of nature and a sense of malice, Christianity forcefully promoted anthropocentrism. Towards the end of the Roman Empire, Christianity had deeply permeated the Mediterranean world. In AD sixth century, the eastern Roman emperor Justinian undertook major civil engineering projects, beginning with the church at Ayasofyain Constantinople

Fig. 9: The Mosaic of Christian emperor Justinian (Sent Vitale, Ravenna, Italy. Photo by Yasuda)

(Istanbul) (Fig. 9) and building vast underground water tanks to guarantee drinking water. Two Medusas have since been discovered from under the pillars of the subterranean vat, the furthest to the back (Figs. 10, 11). One is in profile, and the other is upside-down, forming the base of the pillars. Why however should a Medusa be turned upside-down and placed beneath deep water far underground?

The Christian emperor Justinian (Fig. 9) drove out barbarians such as the Vandals and the Visigoths, at the same time, reclaiming the former lands of the Roman Empire and forcefully persecuting other religions. The Medusa, sunk deep below the pillars at the bottom of the subterranean vats, is a manifestation of this religious persecution. By sinking the Medusa deep underground, he shut her away, never to grace the surface of the world again, and this is reinforced by her being placed upside-down as a symbol of the persecuted.

When the famous Japanese historian K. Abe (1988) was in an orphanage, a snake appeared in the garden. One of the nuns told him, 'It's okay to kill snakes because they're a manifestation of the devil,' and taking up a stick herself, chased the snake around the garden and beat it to death. Abe writes that even as a child, he found this an unacceptable concept.

Fig. 10: Upside-down Medusa under the pillar of the subterranean vat in Istanbul, Turkey (Photo by Yasuda)

Fig. 11: Profile Medusa under the pillar of the subterranean vat in Istanbul, Turkey (Photo by Yasuda)

Fig. 12: For Christianity, the snake is a symbol of heresy. Mosaic of the fighting Jesus Christ (Archbishop's Chapel, Ravenna, Italy. Photo by Yasuda)

For Christianity, the snake is a devil to be scorned, and the Medusa, who had snakes for hair, is a symbol of heathenism (Fig. 12). That is why Medusa was shut away in the deep subterranean darkness while Christianity spread its tentacles throughout the world. Medusa's pathetic voice, persecuted and killed by the armies of Christianity, can still be heard welling up from the subterranean depths.

■ MEDUSA AS MONSTER

No matter how much the missionaries opened up the great forests of north-west Europe, huge tracts of forested land still remained in medieval

Europe. Within the gloom of these forests, the descendants of Medusa lived the remainder of their lives and were termed witches.

During the modern Renaissance from the fifteenth century onwards, however, the forests of Europe were felled mercilessly, and alongside the great flowering of anthropocentrism and modern rationalism, the spirit of Medusa was completely lost.

In the Uffizzi National Gallery in Florence, hangs the artist Caravaggio's (1573–1610) Medusa (Fig. 13). Her head has been completely severed, and is overflowing with fresh blood. Medusa's mouth is open wide, and she is trying to scream. Atop her severed head, the snakes still writhe. The Medusa painted by Rubens (1577–1640), a contemporary of Caravaggio, who is said to have been a devout Christian, is

Fig. 13: Caravaggio's (1573–1610) Medusa (Uffizzi National Gallery in Florence, Italy. Photo by Yasuda)

Fig. 14: P. P. Rubens' (1577–1640) Medusa (Das Kunst Historische in Wien)

Fig. 15: Dense European oak and beech forest (Sababurg, Germany. Photo by Yasuda)

Fig. 16: Deciduous oak with mistletoe (Wien, Austria. Photo by Yasuda)

even more astonishing (Fig. 14). Her head, which has rolled to the floor, is pumping fresh blood, and her wide-open eyes seem about to come flying out at the shock of being decapitated. Such eerie paintings leave a powerful impression on the viewer.

Realism, with its background in modern anthropocentrism and Christianity, depicted Medusa as the ultimate monster. There is no sense whatsoever here of fear at the spiritual power of the snake: only the horror, ugliness, and the unpleasantness of the snake are emphasised.

It was at this time that the spiritual power of Medusa was completely exorcised. Turned upside-down, bullied, and even taken to be a witch, the spirit of Medusa survived. With the raising of the curtain on modern European civilisation, however, Medusa's power was totally destroyed. This meant the complete human domination of the earth. Ever since, the forests of Europe have been completely decimated, and this destruction has spread to encompass the globe. Animals lost their spiritual power, and their mass extinction began. Humans, having lost their fear of nature went on a reckless, apparently unstoppable spree of destruction.

■ 'NO CIVILISATION IN THE FOREST'

It was the Christian missionaries who opened up the gloom of the forests and brought the light of civilisation to the great landmass of Europe. Within the forests lived the Celts and Germanic peoples who revered the great oak tree as holy (Fig. 15). The Druids (priests who governed Celtic faith), attired in pure white robes, would climb the sacred oak at the beginning of spring, and distribute cuttings of mistletoe, the parasite that grows on oak branches. During the terrible winters of continental Europe, when even the great oak shed its leaves and appeared to have given up on life, the evergreen mistletoe alone would grow luxurious, verdant leaves.

The Druids thought that the life of the sacred

oak was gathered and concentrated in the mistletoe (Fig. 16). People would receive small branches of mistletoe from the Druids, and take them home, in the belief that they promised health and wealth; that the mistletoe was a golden wand bringing them happiness. The Christian missionaries, however, could not tolerate the existence of the Druids. 'There are no gods in the forest. There is no civilisation in the forest. It does not matter how much of the forest you destroy if it is for the happiness of the people.' The Druids, who fought desperately to save the great sacred oak, were driven out by the Christian missionaries screaming this mantra, and the pure white cloaks were dyed red with blood (Yasuda, 1991).

Discovery of the Forest Civilisation

■ YANGTZE RIVER CIVILISATION WAS BORN IN THE FOREST

Ever since, humankind has believed, down through the ages, that there is no civilisation in the forest, and that the forest is the enemy of civilisation. Even according to modern history, which has been influenced by the Christian view of the world, the forest is synonymous with underdevelopment and barbarism. Contrary to the Christian view of world civilisations, we found evidence of civilisations that were born in forests: the Yangtze River civilisation in China and the Jomon civilisation in Japan. This was the result of joint Chinese–Japanese academic research, and is ongoing.

The Lomgmagucheng Baodun site is located on the alluvial plain of Minjiang (Min River) (Fig. 17), one of the tributaries of the Yangtze River, of the Schichuwan Province, China. The site is surrounded by a rectangular wall (Figs. 18, 19), the long axis of which is 1,100 m and short axis is 600 m. The total area is over 660,000 m². Radiocarbon dating indicates that the age of this site is over 4,500 yr. BP.

The environmental archaeologists established

Fig. 17: Spring landscape of the alluvial plain in Min River, Schichuwan Province, China (Photo by Yasuda)

Fig. 18: The Longmagucheng Baodung site, Schichuwan Province, China. The wall surrounding the site can be seen in the far distance forming the tree line (Photo by Yasuda)

Fig. 19: Cross section of the rectangular wall of the Longmagucheng Baodung site, Schichuwan Province, China (Photo by Yasuda)

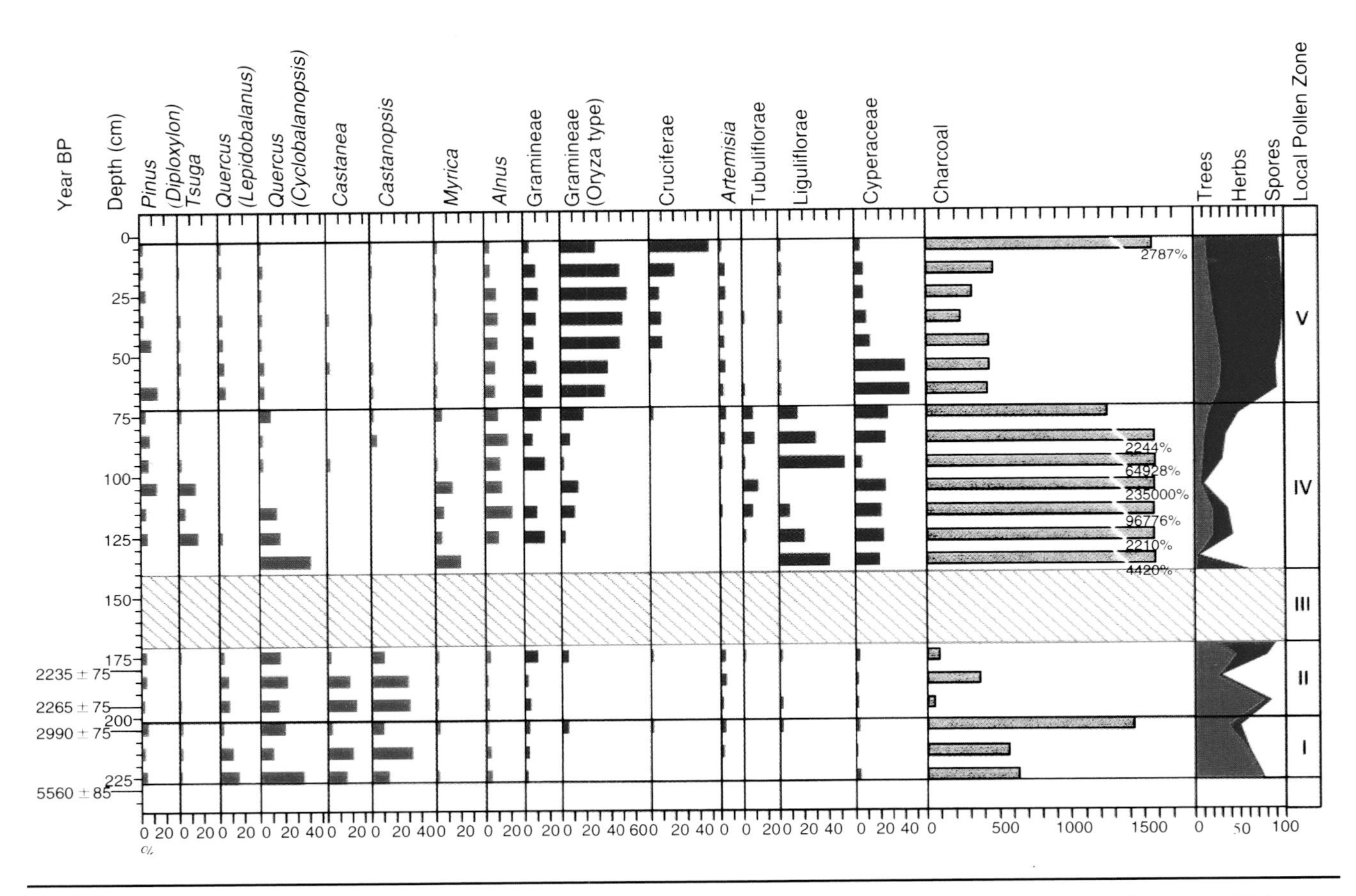

Fig. 20: Relative percentage pollen diagram from the Longmagucheng Baodung site, Schichuwan Province, China

trenches at seven locations within the castle at the Longmagucheng Baodun site and acquired data to determine its age and also to analyse minute fossils such as pollen, diatoms, parasites, plant opal, and large plant and animal fossils. The results of the pollen analysis (Fig. 20) reveal large quantities of pollen of evergreen broad-leaved trees such as oak and chinquapin in the layer of 5,560±85 ^{14}C years BP. In the calendar year, deep evergreen broad-leaved trees flourished in the fan of the Minjiang (Min River) in the vicinity of Longmagucheng Baodun site 6,300 cal. years ago.

This forest was dark and damp, just like the evergreen broad-leaved forests that grow in western Japan. The climate and topography that give birth to the Yangtze River civilisation were the laurel forests that reach as far as western Japan. The laurel forest cultural theory (Nakao, 1967 and Sasaki, 1982) is widely known in Japan.

Downstream of the Tigris and Euphrates Rivers, which gave birth to Mesopotamian civilisation, close to the banks of both rivers there are forests of date palms, poplars, and willows. Away from the riverside, however, there are vast tracts of prairie with little forest. Downstream of the Nile River too, which gave birth to Egyptian civilisation, the forests only exist along the Nile valley. A little further away, the land quickly becomes desert. The basins of the Indus and Yellow Rivers too are belts of dried earth, and there are no deep forests of the kind found in the Yangtze basin. The ravines of dried forests and prairie formed the topography that gave birth to these ancient civilisations.

The Yangtze River civilisation, however, was born in the very midst of a deep, damp, dank forest, completely shattering the general perception that forests do not give rise to civilisations, yet paradoxically it was this very civilisation that completely destroyed the forest, that gave it birth.

The results of the pollen analysis from the area of the Yangtze River civilisation show a reduction in evergreen oak and chinquapin pollen, and a rapid increase instead in grass pollens such as polygonum and mugwort, and fern spores. Moreover, the levels of pollens are thought to denote a rise in the existence of roadside trees and orchards such as willow, plum, and oranges. Also, mulberry pollens become evident, indicating the existence of mulberry fields.

The results of these pollen analyses show the destruction of evergreen oak and chinquapin forests that have been abundant, alongside the construction of castle and town such as the Longmagucheng Baodun site. People obviously felled the large trees of the forest, to build palaces and temples. In addition, they cultivated plums, peaches, and oranges, and planted willows along the roadsides.

What is extremely interesting is that when the Longmagucheng Baodun site was abandoned approximately 3,200 years ago, forests of evergreen oak and especially chinquapin made a temporary recovery. In 2,990 ± 75 ^{14}C years BP (3,250 calendar years BP), the amount of charcoal fragments suddenly dropped sharply, and traces that speak of human activity vanished from the neighbourhood. Also, a layer of silt accumulated that includes almost no fossils of pollen and plant opal. This layer of silt rarely includes pollens belonging to hemlock (*Tsuga*) and spruce (*Picea*), which only grow in the Himalayan foothills (Fig. 21), showing that this layer of silt was carried by floods from the Himalayan mountains and piled up in a short span of time. The question of why the Longmagucheng Baodun site was abandoned can in all probability can be answered by attribution to massive flooding of the Minjiang downstream of the Himalayan foothills.

Fig. 21: Beautiful lake and forest in the foothills of the Himalaya (Jiouzhaigou, Schichuwan Province, China. Photo by Yasuda)

It was from the Han Dynasty (202 BC–AD 220) onwards that the landscape changed into the arable land across which the paddy fields stretch today. Once the Han Dynasty established itself, large-scale cultivation was renewed. This is indicated by the rapid increase in charcoal fragments. With the establishment of the dynasty, we detect charcoal fragments not present earlier on a large scale, which speaks of people living in the vicinity of the Longmagucheng Baodun site once again.

Thus the Longmagucheng Baodun site was redeveloped during the Han Dynasty period onwards, and it appears that the topography and plant life changed greatly. The temple and cemetery, which were on slightly elevated ground, were also destroyed on a large scale during the reign of the Han Dynasty and the ground levelled, and in addition, paddy fields were developed within the castle.

■ COSMOLOGY OF THE FOREST CIVILISATION

The majority of ancient civilisations were born in regions of comparatively sparse forest. Contrary to this, the results of the pollen

Fig. 22: Animal gods with large eyes carved on a ritual jade discovered from Fanshan site, Zhejiang Province, China (after Liangzhu Culture Jade, Cultural Relics Publishing House 1989)

Fig.23: Bronze face mask with protruding eyes from the Sanxingdui site, Schichuwan Province, China (Photo by Yasuda)

analysis testify that the Yangtze River civilisation was born in the dense evergreen broad-leaved laurel forest. The discovery that the Yangtze River civilisation, which made rice production a basic occupation, was a forest civilisation, is extremely significant given the cosmology of that civilisation.

It goes without saying that the forest produces young buds in the spring, overflows with young foliage in the summer, is coloured red in autumn with the dancing russet leaves, and in winter, these leaves fall in cold blasts and the clumps of trees stand rustling. Life repeats its cycle of life and rebirth in the forests. What typifies the cosmology of people who live in the midst of such forests is the cycle of life and rebirth. In the same way that the cycle of life and rebirth repeats itself endlessly in the forest with the visitation of the seasons, humans take birth and die, to be born again in spring. I think that this is the cosmology that forms the basis of a forest civilisation.

What symbolise the cosmology of the cycle of life and rebirth in these forest civilisations are the sun and eyes. A very detailed animal god with a human face carved into a ritual jade tube at 5,000 ^{14}C years BP, which was discovered from the Fanshan site among the Liangzhu cluster of sites in Zhejiang Province, China, proved to be the starting point for the discovery of the Yangtze River civilisation (Fig. 22). What is special about this animal-bodied, human-faced god is that it has two huge bestial eyes. A human wearing a winged hat rides astride the beast, and is touching the huge animal eyes with both its hands. The mouth of this huge-eyed beast has fangs on both sides. Probably, this is a warped version of a Yangtze crocodile.

The most important part of this human–animal hybrid are the two huge bulging eyeballs. Several other examples of more abstractly warped human–animal hybrids have been found, but the one consistent thing about them is the eyes. The core of this animal with a human face can be said to be the two huge eyes.

The bronze ware subsequently discovered 1,500 years later at the Sanxingdui site in Sichuwan Province during 3,200 to 3,700 ^{14}C years BP is also eerie. There are several bronze masks with no torso, only the face (Fig. 23). Moreover, the eyes of these masks protrude abnormally, and are particularly emphasised, and the result is very eerie. The Chinese geographical work *Shanhaijing* relates the legend of eyes that bulged when the sun rose and settled back once more when it set.

These face masks with protruding eyes unearthed from the Sanxingdui site remind me of the statues of the Medusa. As stated earlier, the statues of the Medusa too depicted her body and her feathers in the early period, but when Greek civilisation was at its height they were reduced to depictions of the head alone. In addition, Medusa's eyes also possessed the terrifying ability to turn anyone who looked into them into stone.

■ MEDUSA AND JOMON CLAY FIGURINES (*DOGU*): EYES ARE THE WELLSPRING OF LIFE

In 1994, Aomori Prefecture, Japan was all-abuzz with one of the greatest discoveries in the history of archaeology: the Sannaimaruyama site, a huge village dating from the Middle Jomon Period at 5,700 cal. years BP. Discovered here are architectural remains that used enormous trees that bore great chestnuts with diameters of 80 cm or greater, huge dugouts over 30 m in length, and mountains of pottery. These discoveries completely transformed people's perceptions of the Jomon Period.

Fig. 24: Jomon dogu, 33.3 cm in height emphasising the breast, lower abdomen, genitalia, and eyes (Utetsu Site, Aomori Prefecture. Photo by Tsuboi et al., 1977)

Fig. 25: Light-shielding Jomon dogu 20.2 cm in height with huge eyes (Yoka-machi, Aomori Prefecture, Japan. Photo by Tsuboi et al., 1977)

The north-eastern region of Japan, which was viewed culturally as a remote region from the Yayoi Period (350 BC–AD 300) onwards, actually developed a highly sophisticated culture during the Jomon Period. What characterises the artefacts unearthed from the Sannaimaruyama site, in addition to the countless potteries, is the large numbers of clay figurines. Of these, the board-shaped figurine from the middle Jomon Period was particularly impressive.

With its mouth opened wide as if screaming something, what could this clay figurine have possibly wanted to say? Within its expression of misery, it strikes me that the area around the eyes is particularly emphasised as it is notched. The Jomon Period figurines *dogu* generally emphasise the parts relating to childbirth, such as the woman's breasts, lower abdomen, and genitalia, but in this instance, the eyes are equally emphasised (Fig. 24).

Around the time when the Sanxingdui civilisation developed, statues were also being made in Japan that especially emphasised the eyes (Fig. 25). These are the light-shielding figurines of the late and latest Jomon Periods, and fabricated mostly in the north-eastern region of Japan. Although the Sannaimaruyama *dogu* figurines of the middle Jomon Period already show great interest in the eyes, it was after the late and latest Jomon Periods that special emphasis was placed on the eyes; indeed to such a degree that they resemble dragonfly eyes. Large numbers of these light-shielding *dogu* figurines have been discovered in Aomori Prefecture, at the Kamegaoka site and the Hachinohe City Korekawa site. It is fascinating that the era when these *dogu* figurines with highly emphasised eyes were manufactured in great numbers corresponds to the time when the Sanxingdui civilisation in China was developed, and the Medusa statues were being made in Greece.

From the fact that these huge eyes resemble Eskimo's snow goggles, archaeologists named the clay figurines 'Light-shielding figurines'. What gave rise to this notion was the so-called 'materialist view' of history. At that time, a view of history prevailed that only objects that actually exist could be believed. For a long period of time, it was taboo for Japanese archaeologists to enter the psychology of the Jomon people who made these huge *dogu* figurines.

However, the giant eyes of these *dogu* figurines (Figs. 24, 25) from the late and latest Jomon Periods are not snow goggles or anything of the kind but the final stage expression of an abnormal obsession with the eyes of figurines that was widespread for over 3,000 years from the middle Jomon Period onwards.

Why were the ancient peoples so fixated on the eyes in particular? Also, is there a common thread that links the giant bestial eyes of the animal-bodied, human-faced gods from the Liangzhu site, the protruding eyes on the bronze mask from the Sanxingdui site, the eyes of the Greek Medusa, and the dragonfly eyes of the light-blocking clay figurines from the late and latest Jomon Periods of Japan?

It goes without saying that the eyes are both the weakest part of the human body, yet at the same time the window of communication with the outside world that transmits all external information about that world to the brain. At the same time, when people die, the first thing that happens is the pupils dilate. Today, we debate

Fig. 26: Gift of blue-eye amulets in Turkey (Gift shop in Istanbul. Photo by Yasuda)

whether brain death is true mortal death. In the past, a doctor determined that death had occurred when the pupils were enlarged. The eyes are the organs that control the basis of life. It is fair to say that the human life force is concentrated in the power of the eyes. Ancient peoples also knew from experience that when a person dies, the vibrancy in their eyes disappears. At the same time, they also felt that the eyes were indeed crystalisations of the life force of other people.

In Turkey and Syria, people hate to be stared at. To dispel the evil eye from others, they wear blue-eyed amulets (Fig. 26). When faced with jealousy and loathing from others over, in particular from auspicious events, they will unfailingly wear a blue-eyed pendant to protect themselves. Medusa's eyes had the power to dispel evil in exactly the same way.

Such eyes are the wellspring of life for all living beings, as it is where the life force of other people is concentrated. In all probability this was why the ancient peoples felt afraid of being glared at by other people holding power, ever conscious of the gaze of such powerful Others.

■ EYES TELL OF THE SPIRIT OF THE FOREST

The eyes are the wellspring of life and the ancient peoples were afraid of this mysterious power the eyes possessed, believing them to be the windows to the soul. Besides, they wished for rebirth to a new life through these spiritual windows, and sought to dispel evil through the eye's mysterious power.

The fear they felt towards the power of the eye, which controlled the rebirth of life, drove them to carve the animal-bodied, human-faced hybrids from the Liangzhu site in China, to fashion the bronze mask with the protruding eyes found at Sanxingdui, to make the Jomon clay figurines, the *dogu* with the huge eyes, and to create the eyes of Medusa; which could transform people who looked upon them into stone. In this,

Fig. 27: Beech forest in Mt Daisen, Tottori, Japan (Photo by Yasuda)

the peoples of China and Japan shared a common cosmology regarding the mysterious power of the eyes.

The eyes, which governed the cycle of life and rebirth, were also the windows to the spirit of the forest. The desire for humans to be reborn after death in the same way as the life of the forest, gave rise to great faith in the eyes, creating the huge eyes of the clay figurines and the bronze mask with protruding eyes. The huge eyes of the clay figurines and of the human–animal hybrid gods, which stare at us unblinkingly, tell of the spirit of the forest (Fig. 27).

I believe that this is deeply connected to the fact that people were living surrounded by forest. Japan's Jomon Period goes without saying, but the people of the Liangzhu site in Zhejiang Province, China, as well as the people who made the mask with the protruding eyes at Sanxingdui in Schichuwan province, China, and the ancient people of Greece and Turkey, who created the Medusa, originally lived surrounded by dense forest.

When living surrounded by forest, people felt the gaze of the gods of the earth staring fixedly upon them, so they fashioned statues with such eyes. This unceasing gaze thrown at them from the forest speaks of the endless cycle of life and rebirth of all life in the same way as for life in the forest.

The forest was the abode of the gods of the earth. People were in awe of the spirit of which the forest spoke, and thus created statues with huge eyes. However, the statues, with eyes that stared upon the people, ceased to be made once a certain temporal boundary had been crossed, and in the end, they were destroyed. That temporal boundary coincides with the time when the forests were destroyed.

When the bronze mask from Sanxingdui was destroyed and burned, when the Jomon clay figurines ceased to be made, when Medusa's head was sent tumbling from the beams of the temple, the forests disappeared or incurred severe damage. When the forests disappeared and thus lost their spirit, people stopped making the huge-eyed statues that stared upon them.

I feel that when people stopped making giant-eyed statues, an era came to an end. It was the end of the era of the spirit of the forest civilisation; the end of the era of animism.

Production of the giant-eyed clay figurines that had continued to be made for over 3,000 years during the Jomon Period suddenly stopped with the beginning of the Yayoi Period. The only way this can be viewed is that as the background to this, there was a fundamental change in the relationship between Japanese people and the forest.

The raising of the curtain on the Yayoi Period brought the onset of large-scale deforestation. Amidst the expansion of paddy fields and villages, the lowland forests were destroyed. With this advancing deforestation, the spirit of the forest embraced by the Jomon people was gradually lost. The Japanese forests changed into secondary woodlands close to human settlements, recapitulating once more the spirit of the forests of the Jomon Period, but this time there were no people making huge-eyed statues. In the same way, people stopped making these huge-eyed statues after the destruction of Sanxingdui civilisation. In both these cases, the spirit of the forest that led to the creation of huge-eyed statues was lost forever to the souls of the Japanese and Chinese peoples.

Both Japan and China still stretch out today the world of animism and the spirit of the forest. The world of animism in particular created the foundation for the spiritual cosmology of East Asian culture, and is profoundly linked to the spiritual cosmology of both China and Japan. When the peoples of Japan and China rediscover in a genuine sense their links to the world of the forest and animism, East Asia will create a new and great cultural and civilisational ethos for the twenty-first century.

References

Abe, K. (1988): The discovery of the history in my life (*Jibun no Naka ni Rekishi wo Yomu*). Chikuma Shobo, Tokyo, 204pp.

Fukube, N. (1987): Journey of Greek Arts (*Girisha Bijyutsu Kiko*), Jiji Tsushinsha, Tokyo, 352pp.

Nakao, S. (1967): Origin of Agriculture (*Nogyo-Kigenron*). Morishita, M. and T. Kira (eds.), Nature-ecological study (*Shizen Seitaigakuteki kenkyu*). Cyuokouron-sha, Tokyo, 329–494pp.

Sasaki, K. (1982): The Route of Laurel Forest Culture (*Shoyojyurin Bunka no Michi*), NHK Books, Tokyo, 253pp.

Tsuboi, K. et al., (eds.) (1977): Archaeological Treasuries of Japan 3- Clay figurines, Haniwa. Kodan-sha, Tokyo, 219pp.

Yasuda, Y. (1991): The Age of Mother Goddess (*Daichiboshin-no-Jidai*). Kadokawa-shoten, Tokyo, 240pp.

—— (2000): The Birth of Great River Civilizations (*Taigabunmei no Tanjyo*) Kadokawa-shoten, Tokyo, 354pp.

Part III

FOREST AND WITCHES

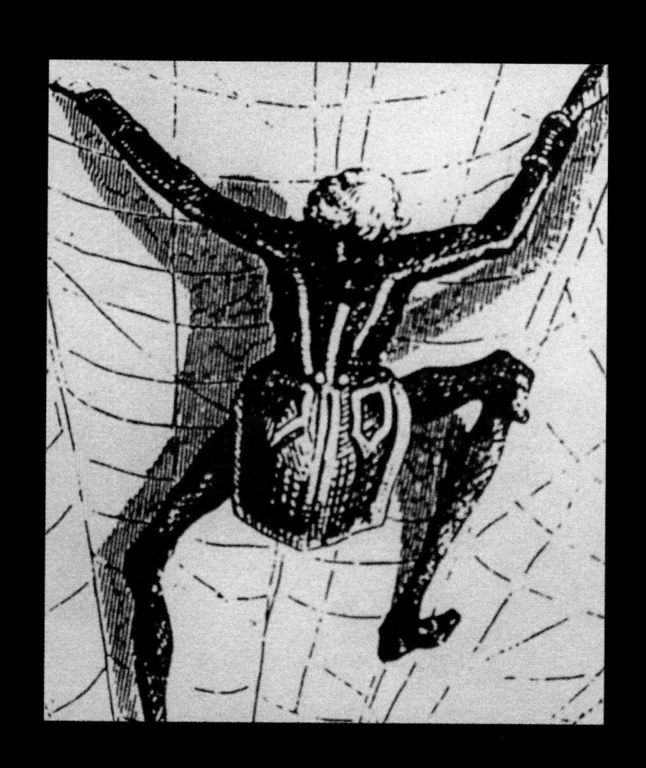

CHAPTER 9

The Origin of the Earth-Mother Cult in Europe and Japan[1]

ATSUHIKO YOSHIDA

The Earth-Mother Cult in Palaeolithic Europe

It is clear that the religion of *Homo sapiens* began with the adoration of the Earth-Mother. She was a grand goddess believed to be the supreme ruler of universal fecundity. The first traces of this religion, the so-called stone or ivory 'Venuses', are found at the beginning of the Palaeolithic period between the Ural chain and the Pyrenees, and are arguably the oldest existing idols. The 'maternal' body parts of these statues (breasts, abdomen, vulvae, and buttocks) are strongly accentuated, while the rest of their bodies are frequently minimised or left out. The artists clearly did not intend the statues to be life-like representations of pregnant women but rather the representation of a supernatural being who accomplished, with her overflowing genital organs, supremely performed the maternal function. This interpretation is supported by the posture of many of the statues, such as the Willendorf 'Venus', whose forearms and hands are folded over her enormous breasts, apparently pressing them for milk, while her head is deeply inclined forward. Although her eyes are not represented, the tilt of her head suggests she is directing a tender look towards her nipples, abdomen, and vulva, so that, in my view, she can be interpreted as an idealised figure of the Earth-Mother performing the threefold maternal functions (breast feeding, pregnancy, and birth). She is clearly supposed to ensure the unceasing renewal of nature, fauna, flora, and the human species.

Upper Palaeolithic art also appears to contain representations of the Earth-Mother. Most of the Upper Palaeolithic paintings and engravings in the Franco–Cantabrian region are in spacious underground galleries situated far from the cave entrance. They are generally very difficult to get to: some corridors, such as those in the Trois Fréres cave in Arriege, are very long (up to a kilometre) and are so small that a human would have to crawl or wriggle through to gain access.

There can be little doubt that Upper Palaeolithic man imagined the long labyrinthine corridor leading to the sacred cult centre as the vaginal path of the Earth-Mother, which he had to traverse in order to enter her uterus and perform ceremonies which he believed to be absolutely vital, not only for his own existence, but also for that of the world as a whole. By painting herds of animals on the walls and ceilings of this sacred space he may have believed that this would ensure ongoing life and a continuing abundance of food. We can also infer with certainty that each time he penetrated the grotto to enter the uterus of the

[1]The original paper was submitted in French and translated by the editorial staff at International Research Center for Japanese Studies. The quotations were originally translated by the author.

Earth-Mother, he would have imagined that he was experiencing symbolic death. Conversely, when he left the cave this would have symbolised his rebirth. Consequently, there can be no doubt that by exiting the cave he would feel reborn as one of the Earth-Mother's numerous children, thus confirming the link between himself and the goddess.

The Earth-Mother Cult in Japan

Evidence of an Earth-Mother cult in Japan at the beginning of the Jomon period (10,000 BP) is provided by female terracotta figurines or Jomon *dogu*. Although the earliest examples (up until about 5,500 BP) are relatively unrefined, it is nonetheless possible to distinguish accentuated genital organs, breasts, abdomen, and vulvae. The earliest examples are very small and simple in form and do not exhibit much variety, but this situation changes in the Middle Jomon period.

At the beginning of the Middle Jomon (ca. 5,500 BP) the number of these figurines suddenly increases and they become much more refined in form: their shapes become more complicated and varied and their size increases—some are over 30 cm long. Almost all of the Jomon *dogu* of this period are found in fragments at archaeological sites, and it has proved impossible to reconstruct these. During field-work at Shakado, in the Yamanashi department, 1,116 Jomon *dogu* fragments were found, none of which were intact or could be reconstructed.

Some of the *dogu* fragments have been discovered under conditions that lead us to clearly infer that they were cult objects. For example, in Toshikura (Department of Nigata), an important site of the Middle Jomon era, a 51 cm deep and 26 cm diameter hole had been dug into the floor of a building. This hole had been filled with black soil and on top of it the torso of a figurine, whose head, both arms, and the lower parts of the body below the hips had been broken off, had been erected at the centre using pottery fragments as support. These fragments were also surrounded by eight other pottery fragments painted in red which had been placed against the wall of the hole.

It appears that the figurines were broken into pieces, each of which symbolised a part of the goddess's body, which were then taken to different places. Part of the goddess was left in the house to receive adoration. It also appears that at the beginning of the Middle Jomon period, perception of the Earth-Mother changed and it was thought necessary to murder her in some way by destroying these figurines. However, it was also believed that the vital force of the Earth-Mother was so infinite that despite being symbolically destroyed, she continued to provide all the natural resources that humans needed to survive

In my view, it can be inferred that it was the beginning of agriculture that provoked the start of this remarkable religion. A few Japanese archaeologists have already observed that at the beginning of the Middle Jomon period, the important change in the cultural domain was the result of the introduction of agriculture. One of the archaeologists, Teruya Ezaka, has posited that the principal cultivated plants were taro and yam, firstly because of the sudden appearance in the archaeological record of a stone tool that can only be used for digging soft earth (a large number of these tools, which seem particularly appropriate for planting and harvesting tubers, continue to be made and used in Japan), and, secondly, because of the invention of a pot that appears to be suitable for cooking tubers. This type of pot was made in large numbers during the Middle Jomon period.

Thus, it can be inferred that the population of the Middle Jomon, which was just beginning to grow tubers, felt that the Earth Mother's body was injured each time the soil was worked. Support for this view is forthcoming in the following quotation from Sumoholla, an indigenous North American prophet from the

Umatilla tribe, who expressed his disgust for agriculture at the end of the last century. Eliade (1957: 207) has used the words of Sumoholla to demonstrate the universality of the concept of a relationship between the Earth-Mother and agricultural work:

> It is a sin to wound or to cut, to rip out or claw our communal mother with agricultural work. You ask me to plough the soil? Will I go and take my knife and stab the lap of my mother? But, then, when I am dead, she will not take me back to her lap. You ask me to dig and remove the stones? Will I go and mutilate her flesh in order to get to her bones? But, then, I will not be able to enter her body any more and be born again. You ask me to cut grass and hay, and to sell it, to get rich as the white man? But, how will I dare to cut the mother's hair.

The men of the Middle Jomon period adopted agriculture but simultaneously experienced feelings similar to those of Sumoholla, so it was natural for them to believe that the Earth-Mother had been injured and mistreated by their labour. However, in spite of such injury, her benevolence was such that she did not stop providing the resources necessary for human survival. In Japanese mythology, found in the *Kojiki* and in the *Nihon Shoki*, at the beginning of AD eighth century there are many stories that appear to testify to such a belief. The following paragraph is part of the contents of one of the myths found in the *Kojiki*:

> A very impetuous god, named Susano, who is one of the protagonists of Japanese mythology, went to see a goddess whose name, Ogetsuhime, literally means 'Great Princess of the Food', and asked her for some food. She responded to his request by offering him a variety of dishes made from all sort of things which are good to eat, and those she obtained from her nostrils, her mouth and her anus. Susano, who had spied on her, sure that the goddess was making him eat litter, lost his temper and killed her. A silkworm appeared from her head, rice from her eyes, milk from her ears, *Phasoleus radiatus* (a kind of red bean) from her nose, wheat from her vulva, and soya from her anus. This food was picked up and cultivated by a divinity, called Kamimusubi.

Another story which, in spite of the enormous divergence, obviously belongs to the same kind of myth as the previous one, is found in the *Nihon Shoki*:

> One day, the sun goddess Amaterasu asked the moon god, Tsukiyomi, to go down on Earth and to visit a divinity called Ukemochi ('Holder of the Food'). When Tsukiyomi introduced himself to her, Ukemoshi vomited rice while she was facing the plains, then fish, large and small, while she was facing the sea, and finally soft-haired and hard-haired game while she was facing the mountains, and she wanted to make a feast for the moon god with the well-prepared food that was on the huge tables. Upset by the behaviour of the goddess, which he believed to be rude and humiliating, the red-angered Tsukiyomi killed Ukemochi with his sword and, back in the sky, recounted to Amaterasu a detailed report of his mission. Tsukiyomi's violent acts roused Amaterasu to indignation to such a degree that she declared she never wished to see him ever again. Since then, the sun and the moon never appear together in the sky.
>
> Amaterasu then dispatched a god named Amenokumahito ('Celestial Man–Bear') to the murder scene. On the different parts of her body, he found the horse, the beef, the silkworm cocoon, and six other cereal species (the five species mentioned in the previous myth plus the panic) and he took away with him all the food to the sky. Amaterasu was very pleased and said: 'Here is what men will have to feed from in order to live'. And she established agriculture and threading in the sky.

There is no doubt that these two stories of the *Kojiki* and *Nihon Shoki* are very similar to the well known myth of Hainuwele (Jensen, 1966; Yoshida, 1966) narrated by the Wemale from the island of Ceram in Indonesia. The following is a brief summary of this myth, which is in actuality very long and detailed:

> The blood that a man named Ameta poured on to flowers of a coconut tree changed into a child named Hainuwele ('Coconut Spray') by mixing with its nectar. Hainuwele was not an ordinary human; she grew at an incredible speed (after three days, she was already nubile), and her excrement was very sought after and expensive, like Chinese plates and gongs, so Ameta became very rich. When people from the village organised the dance Maro [this dance lasts nine nights, and, while the men dance the whole night, the women, sitting in the middle of a circle in a spiral formed by the dancers, give them things to eat], Hainuwele stood in the centre and distributed, not only to the dancers but also to the spectators, some wonderful products from her body. The maid, having distributed eight times during the dance, gives more and more precious things each time. The inexhaustible wealth finally provoked jealousy and repugnance such that on the ninth night, men killed her by burying her alive and by dancing on top of her body. Her body was recovered the following morning by Ameta, who, after having torn her body into pieces, buried them all over the dancing ground. From these pieces, strange things unknown to the world were born, in particular, several yam (*Ainte*) and taro (*Uku*) species; *Ainte latu paite* from her limbs, *Ainte babau* from her breasts, *Ainte mau* from her eyes, *Ainte moni* from her sex, *Ainte* ka *oku* from her hips, *Ainte leleila* from her ears, *Ainte jasane* from her feet, *Ainte wabubua* from her thighs and from her head *Uku joijone*, etc.

Apart from this important myth, Jensen (1939: 69) collected from the Wemale the following story which is also about the origin of cultivation,

> A grandmother living with her grandson used to give every day a pap that she had prepared while he was away. One day the boy was spying on his grandmother, and he discovered that she made the good dish from the filth that came off her body. When she called him, he said, 'I will not eat because I saw what you were doing'. Then the grandmother answered, 'If you have seen and if you do not want to eat, go away. But you must come back in three days and look at the house. You will find there something. When the child came back he found palm trees, which had grown from her corpse, instead of the grandmother: a tree of Areng from her head, a coconut tree from her sex, and a few other Sago trees from her body. By their trunks, there was also a stick for planting and other agricultural tools.

Such myths belong to the same type as those that are distributed throughout Japan, Indonesia, Melanesia, Polynesia, and also America. In some places, as in the case of Japanese mythology, the origin of cereals (rice in Indonesia and corn in America) is explained by this kind of story (Hatt, 1951). However, as has been convincingly argued by A.E. Jensen (1966:164), the myth in question seems to have originated from what Jensen has termed the '*die Altpflanzer*', or sometimes the '*die fruhen Pflanzervolkern*', culture.

The economy of the 'old cultivators', which can nowadays be found in tropical regions such as Indonesia, Melanesia, Polynesia and Africa, relies essentially on the cultivation of taro and yam in fields cleared by setting fire to the undergrowth and fruit picking from trees. According to Jensen, the cultivation of taro and yam constitutes the most ancient form of agriculture in the world.

In addition, Jensen has shown that one of the most prominent characteristics of the culture of these *fruhen Pflanzervolkerm* is the performance of sometimes violent sacrifice, in which can be recognised dramatic representations of the primordial events related in the myths set out above. During these ceremonies, the victims were killed and some of their body parts were eaten. What remained of the corpse was then divided and shared so that anyone could bury them in or spread pieces of them over their plantations. For example, at the beginning of the final part of an important feast called Moguru, the Kiwai of New Guinea captured a wild boar which, after having been decorated in a sumptuous fashion, was then solemnly immolated on an especially built altar in front of the central pillar of a cultural house (the Darimo). The animal was then cooked by an old woman, who was the only woman allowed to enter this area. Only the elders had the right to eat the meat. The remains of the wild boar were then buried in fields where tubers or palms were cultivated. The Majo and Rapa feasts organised by the Marind-amin of the same island were notorious for their ferocity: after a young woman was gang-raped, she was then killed and eaten, her remains subsequently treated in a similar manner to those of the wild boar of the Kiwai. There can be little doubt that by treating their victims in this way, the tribes of ancient cultivators were in some way reproducing the mythical events which they believed had ended the primordial age: i.e. the killing and dismemberment of the Earth-Mother.

However, the agriculture which began to be practiced in the Middle Jomon period is likely to have been similar to the *fruhen Pflanzervolkern*. We have seen that this agriculture consisted of taro and yam cultivation like that of the ancient farmers. On the other hand, it is also clear that nuts such as chestnuts, walnuts, and acorns, had also been an important food source for the Jomon people since the beginning of this period and that, during the Middle Jomon period, the propagation of these fruit trees began in order to cultivate woods in proximity to populated areas.

It can be believed that the people of the Jomon period explained the origin of the tubers they cultivated, and maybe also the origin of certain trees (such as the chestnut tree, the hazelnut, and the oak), by a myth similar to that of the Wemale, and that the breaking of the Earth-Mother figurines into fragments has a ritual meaning which reproduced the killing and dismemberment of the goddess as related in that myth. That myth could then have gradually transformed itself into the type of myth documented in the *Kojiki* and the *Nihon Shoki*, as the cultivation of cereals, and the farming of the silkworm and livestock was successively introduced into, and spread throughout, Japan.

In addition, there is no doubt that Middle Jomon men portrayed the benevolence of the goddess in their pottery. The large vases that they ordinarily used as kitchen appliances often have a decoration in the form of a head at the edge of the opening edge, whose face is identical to those found on the figurines. Also, as the trunk of the vase is usually clearly inflated in its middle part, there is no doubt that the pieces of pottery as a whole was indeed a statue impersonating the pregnant goddess. It is also clear that when the Jomon people were preparing dishes in these pots, they imagined that the food was produced in the goddess's body, as represented by the vase.

So, from the existence of this type of pottery, one can infer with certainty that Middle Jomon people, on one hand, and eighth century Japanese, on the other—the *fruhen Planzervolkern*—already related in myth that the goddess, whose killing and dismemberment at the end of the primordial age had given birth to nutritional plants and other crucial resources, also had the miraculous gift of providing food in abundance by excreting it from her living body.

References

Eliade, M. (1957): *Myths and Mysteries (Mythes, rêves et mystères)*, Librairie Gallimard, Paris, p. 207.

Hatt, G. (1951): The Corn Mother in America and Indonesia, *Anthropos*, 46: 853–914.

Jensen, A.E. (1966): *Die getötete Gottheit: Weltbild einer frühen Kultur*, W. Kohlkammer Verlag, Stuttgart·Berlin·Köln·Mainz, 164 pp.

—— (1939): *Hainuwele: Volkserzählungen von der Molukken-Insel Ceram*, Vittorio Klostermann, Frankfurt am Main, p. 69.

Yoshida, A. (1966): The Excretions of the Goddess at the Origin of Agriculture (Les excrétions de la Déesse et l'oigine de l'agriculture), *Annales (Economies, Sociétés, Civilisations)*, 39: 717-28.

The Didima Temple in Turkey (Photo by Yasuda)

CHAPTER 10

European Forests, Fairies and Witches in Medieval Folklore[1]

PHILIPPE WALTER

During the Middle Ages, the European forest was a hostile and frightening place. The etymology of the French word *sauvage* (savage in English) is proof of this: the word derives from the Latin *silvaticus* meaning 'pertaining to the forest' or 'an inhabitant of the forest". In fact, the 'savage', or 'wild man of the woods' only comes from the forest. The savage and the sylvestrian are two aspects of the same cultural entity, imbued with sacred dread.

Above all, the forest was perceived as a place where a variety of dangers were encountered, particularly because it was traditionally the dwelling place of wild men and women. They are however, merely the reincarnation of the ancient deities of the woods, the mother-goddesses of antiquity, which the Middle Ages transformed into fairies. Thus, the forest is above all a place of fairy magic, a place of adventure and wondrous feats. It is in the forest that a mystical dialogue is established with the deities of nature who promote bounteous harvests. It is the favoured home of spirits who entice human beings into their haunts and it is a place of mystic rites, as is clearly shown by its common identification as a haunted place.

With the advent of Christianity, a religion of the Word (founded on Biblical text) replaced the ancient religion of nature. From this time onwards, the relationship between Nature and the Divine radically changed. Whenever places of Christian worship were built on ancient pagan sites, such as fountains, the mythology previously attached to the ancient rites was fundamentally transformed (Walter, 1992). The old pagan tales became folklore beliefs, and fairy magic frequently changed into witchcraft: places that had previously been thought of as the abode of fairies became instead the sites of a witches' sabbath. Along with the Magic Mountain, the Enchanted Forest now harboured witches who performed strange demonic rites.

As the French historian Jules Michelet (1866) showed in the nineteenth century, witchcraft appears to be a form of cultural resistance to the values of 'modern' society which tends to sever the ancestral link between man and nature. The witch remains identified with the primitive rites and behavioural patterns that formed part of the superstitious mentality of the pre-Christian world. As Delumeau (1978: 450–473) has observed, the real cultural explosion in witchcraft did not correspond, as is often thought, to the medieval period, but rather to the sixteenth and seventeenth centuries, when the establishment of the new order created a social crisis. In fact, the 'golden age' of witchcraft corresponds to the birth of Cartesianism and to the idea of social

[1]Translated by Robert Griffiths.

development through exploitation of the natural world in the developing industrial era. Witchcraft is the affirmation of a civilisation in which the centuries-old pact between man and nature is still intact.

The Great Celtic Woods

Almost by definition, the Celtic and Medieval forest was a haunted place; never an empty place. For the ancient Celts, there was a semantic correlation between 'wood' and 'sacredness': every Celtic forest was principally a sanctuary. It was the *nemeton* or 'sanctuary-forest'; a place where Druids lived and officiated (Guyonvarc'h, 1990: 34–6).

A story in Lucan's *Pharsalia* depicts the Druids living deep in the woods (*nemora alta*) and retreating into uninhabited areas. There they practised barbaric rites and a sinister cult. A scholarly explanatory note adds: 'they worship gods in the woods without resorting to temples' (Guyonvarc'h, 1990: 452). Another passage in the same text relates that near Marseilles Caesar cut down a sacred wood whose sculptured tree trunks were 'sad representations of the gods' (Guyonvarc'h, 1990: 452). A Druid who officiated there in honour of the local deity was even said to have carried out human sacrifice. The text reads: 'the priest himself feared to enter the wood since he was in fear of the master of the wood' (Guyonvarc'h, 1990: 339).

The Celtic temple was thus an integral part of the forest itself, and the Druidic sacred rites were established in the midst of trees. This feature of the Celtic world is distinguished from that of the Romans and Greeks in which temples are situated in cities. Moreover, the Celtic religion did not suppose a collective participation in its rites, but reserved its worship for the mediators with the gods. These privileged intercessors carried out divine worship on behalf of their community in the heart of the forests and far from the cities. (Guyonvarc'h, 1986: 228–31).

The special role of the forest was due to its status as the place in which to communicate with the deities. In the fifth century, Martianus Capella counted the invisible beings, who peopled the forests, lakes and rivers:

> Places inaccessible to men are peopled with a whole host of very ancient creatures who inhabit the forests, woods and sylvan sanctuaries, the lakes, springs and rivers; they are called Pans, Fauns, Satyrs, Fonts, Sylvans, Nymphs, Fatui and Fatuae or Fautuae or even Fanae, and from this name is derived that of Fana which refers to their prophecies. All these creatures are mortal but die at a very advanced age; and they possess to the highest degree the power to predict the future, and the power to harm humans (Harf-Lancner, 1984: 20)

IN THE BEGINNING WERE THE FAIRIES

From the outset, forests were considered dangerous places where human lives were at risk from the spirits who sought to subjugate them, and who, with magical powers, held the key to their destiny. The word 'fairy' derives from the Latin word *fata* which means 'destinies' (Harf-Lancner, 1984: 59–60). The fairies were also diviners and able to foresee the future. Is this not already a way of stressing that the forests hold the future of humankind and the hidden truths of life; truths which humankind is constantly striving after? What however are these truths?

In the tenth century, a German Bishop, Burchard of Worms, referred more specifically to the female creatures who were the veritable demons of the forest:

> You have believed what many customarily believe—that there exist wild female creatures, called 'women of the forest', said to be carnal creatures, who, when they so wish, materialise before their lovers and, it is said, take their pleasure with them; and then in the same way, when they so wish, they hide away and vanish [Walter, 1989: 487–8).

These women of the forest resemble succubae. They belong to a demonic world but at the same time they, apparently paradoxically, represent the values of pleasure and love. The Bishop denounces the evil spirits of erotic love because of the renunciation of the flesh which man must seek through God. In his discourse, religious values and those of profane love stand in sharp contrast. Christianity rejects the whole world of profane love as the work of the devil ('the tasty meat of the devil' according to ecclesiastic expression), as it also rejects the whole world of the forest, for the 'women of the forest' represented the forces of eroticism and sexuality. In the Middle Ages, the forest was a metaphor for the forces of creation. It was also depicted as a place of life-giving energy, as a reservoir and a model of human fecundity. It is a domain that was sometimes uncontrollable, but was also a world indispensable for the continuity of life (Evola, 1976).

The old medieval characters sometimes reappear in folklore. The Tchouvakians of the Volga region refer to a male demon called the 'father of the forest' and his wife, the 'mother of the forest'. The latter is a hirsute woman who wanders totally naked, her long pendulous breasts hanging from her shoulders. She tries to attract men by showing them her vulva but when they allow themselves to be seduced, she tickles them to death. The Wends, a Slav tribe living between the Elba and the Oder, also refer to a woman of the forest with pendulous breasts. She is nicknamed the 'madwoman' and has shaggy hair, fiery eyes and lives in a cave. She emerges at around midday and pursues any young men who happen to be in the forest. Like a sphinx she asks them all sorts of questions and if they do not give the right answers, forces them to submit to her caresses and kisses. If they try to escape she grasps them and thrusts her long forked tongue into their mouth (Roheim, 1974: 259). In reality, this woman of the forest is the embodiment of a middle-aged sexual temptress.

In his recent analysis of medieval beliefs about the forest, Lecouteux (1995: 85–107) has stressed the primordial role of place and the 'spirit of place' in the elaboration of myths and legends of the Middle Ages in Europe. He has succeeded in uncovering the hidden issues at stake, when a particular plot of land is appropriated by a human community. Most of the creatures, fairies or demons, revealed through the medieval imagery or literature, are archaic creatures, the 'spirits of the place': invisible or subterranean spirits who hark back to a primitive folk memory and to a quasi-religious respect for nature. The mythological beliefs refer to the fundamental idea of a spirit world which requires the 'conquest and defence of the earth' by human beings. From this belief arise all sorts of conflicts and pacts with the 'spirits of the place' to achieve the best possible conditions for human occupation. The 'women of the forest', the 'mother' or 'father of the forest', are 'spirits of the woods' who make the forest what it is: a primitive and sacred place which humans cannot appropriate with impunity.

In Celtic origin myths, this 'other world' was not a hypothetical place distinct from the terrestrial world but, to the contrary, was a continuum of the everyday world. It could be a forest, a stone or a spring, which marked the frontier between the 'other world' and the ordinary world. A folk story from Lorraine in which all the action takes place in a forest is typical of a whole host of medieval fairy tales (Tobin, 1976; Micha, 1992):

> A shepherd was wandering by a rock named Nonnenfelsen or Teufelsfelsen [Nuns' Rock or Devil's Rock]. He suddenly caught sight of a white stag observing him with a fixed stare. He moved towards the animal, which bounded away among the trees until it reached the Nonnenfelsen. The stag entered the rock, the man followed him and found himself inside an illuminated cave. Three beautiful maidens seated to his right joyfully welcomed him. Looking around, the

shepherd saw a gigantic animal crouched on a large iron chest. The eyes of this horned monster emitted flashing sparks, his nostrils puffed out smoke, and in his mouth he held a golden key. The maidens asked the shepherd to rescue the key from the jaws of the monster so that they might be freed for ever, but the terrified shepherd took flight. In this region it is said that every Easter morning, three maidens come from the wood to bathe in the pure spring water of the valley. Then, they return sadly to the woods for they are forever prisoners of the monster, and no-one since has attempted to liberate them [Printz, 1956]

In this story, the three young maidens are obviously fairies. One of them, transformed into a white doe, has the task of bringing the hero to the heart of the other world through the rock in the forest. It is there that the young man is initiated into a mystery which harks back to the great themes of pre-Christian mythology. The horned monster is certainly the Gallic 'Cernunnos' whose very name seems to be echoed in the name of the rock in which he lives (Nonnenfelsen = Cernunnos). In gallo–roman statuary, this god with the horns of a stag is often depicted with a bag full of coins which he spills forth before a stag or a bull, both horned animals. Sometimes he is depicted with a goddess, holding a horn of plenty. Duval (1993: 47–8) describes him as 'a god of earthly fecundity' for 'he has the strength of the king of the forest and the power of a man who can create wealth: to him can be attributed the bounty of nature, in both its aspects, the wild and the cultivated'.

When Yvain, the knight in the tale by Chrétien de Troyes, sets off on his adventures in the company of a lion, he finds himself in the Breton forest of Brocéliande, referred to by Chateaubriand (1989: 164) as 'the dwelling place of the fairies'. There, Yvain encounters love in the guise of a fairy destined to become his lady. Once again, true love that changes a man's destiny is found in the heart of the forest. However, before finding his fairy, Yvain encounters a mysterious man of the woods who resembles the Cernunnos referred to above. This character is the medieval descendant of a venerable deity of the ancient Celtic forests. The text tells us that he had an enormous head, bigger than that of a warhorse or any other beast, with tousled hair and a pock-marked brow two spans wide, huge hairy ears like those of an elephant, bushy eyebrows, a flat face, the eyes of an owl and the nose of a cat, a mouth like a wolf, the teeth of a boar (all sharp and red), a ginger beard and curled moustache. Leaning on his club, he was wearing strange apparel of neither flax nor wool, and round his neck were draped two freshly skinned hides of oxen or bulls. Chrétien is clearly describing the wild man of the forest (Bernheimer, 1952: 27–8) who holds sway over wild animals, in particular the savage horned beasts which he masters with ease. A gallic text which refers to the same character adds that he ruled over snakes and wild cats. A veritable 'spirit of the place', the wild man rules over the universe of the forest and over all the beings that inhabit it. He is the incarnation of the old deity of destiny who had the power of life and death over all his subjects.

■ THE MYSTICAL FOREST

Because it is rooted in the pre-Christian sacred world, the medieval forest is also a place of spiritual revelation. The life of Saint Hubert, the son of Bertrand, Duke of Aquitaine, is well known. Saint Hubert was so passionately devoted to hunting that he gave himself over entirely to this activity in the forests of the Ardennes, even hunting on Good Friday, the day commemorating the death of Christ, which was considered a great sacrilege. It was on that day, just as he was about to bring down a great stag, that the animal turned to face him. Between its antlers it bore the radiant image of Christ on the cross, with the words 'Oh Hubert, how long will this vain passion make you forget the salvation of your soul?' Thunderstruck, the young prince threw away his hunting dagger and realizing the extent of his errors, sought absolution by leading the life of a monk. Accepting his repentance, Pope Sergius later ordained him as a priest and he eventually became bishop of Liège and of Maastricht.

This exemplary conversion of Saint Hubert was preceded by that of Eustatius Placidus, a Roman who was an enthusiastic hunter living during the reigns of the Emperors Titus and Trajan in the first century after Christ. While hunting in the forest, he too saw a stag turn to face him, bearing the burning image of the crucifixion between its antlers. 'Placidus, why do you persecute me? I long for your salvation.' It is a theme of the hunter being hunted. The fierce Placidus converted to Christianity and died as a martyr in the reign of the Emperor Hadrian.

In this legend, the forest is a ritual place in which spiritual conversion occurs because it is the place of privileged encounter with the deity. This is the Christian version of a pagan fable. In this later version the stag, with the figure of Christ, replaces the spirit of the enchanted forest, which in the Celtic legends often takes the form of a stag or a doe. It is often in the forests that the destiny of exceptional men is forged.

Faced with a wooded universe haunted by mysterious spirits of pagan origin, Christianity endeavours to convert the forest into a desert. As Le Goff (1985: 62–3) explained, 'in the medieval Western world the contrast [was] not between the city and the country as [it was] in Antiquity' but 'the fundamental dualism of culture versus nature is illustrated by the contrast between what is built, cultivated and inhabited (the city–castle–village) and what is naturally wild (sea, forest, and desert plain), by the contrast between human societies and the world of solitude'.

For monks, the charm of the forest can be explained by their desire to withdraw and their rejection of normal society. Saint Bernard, founder of the Cistercian order, declared, 'The forests will teach you more than books. The trees and rocks will teach you things that the masters of erudition cannot'. That is why many Cistercian monasteries were built far from towns and often in the heart of ancient forests, unlike those of the Benedictine order which were usually situated in urban centres.

It is clear that the location of monasteries in the depths of forests was a continuation of the old Druidic religious customs. The scattered evidence of these Christian cult centres still exists in Europe today. The new cult merely Christianised ancient pagan practices. To take control of the Western world, Christianity had to compromise with the native religions that had preceded it. Thus it took over the ancient pagan practices which were still in force and converted them to the 'true God' of the Bible.

The founding of the Grande Chartreuse Order near Grenoble by Saint Bruno and his disciples continued the ancient pagan desire to occupy awe-inspiring places considered sacred in order to establish a special dialogue with the deity. Saint Bruno could thus address his novice monks to show them the reality of the forest at that time,

> You will find a forbidding place, the lair of wild beasts, of towering mountains, of vast forests, an intense and enduring cold, no fruits, no crops. Such is the land to which I lead you. The tumult of the torrents, the silence of the woods; everything heralds death, everything instills fear in that place [Bligny, 1984]

If one accepts that the name of Bruno resembles that of Brun, the name of the bear in the Roman de Renart, it is clear that in Christian and popular mythology, Brun(o) of the Chartreuse Mountain reincarnates the old Celtic myth of the bear and the wild man whom the Chartreuse monks are ultimately there to emulate. Bruno of the Chartreuse Mountain can be compared with Brun of the Mountain referred to in an epic romance of the fourteenth century (Meyer, 1875). Brun is a child, abandoned at birth in the forest of Brocéliande in Britanny, close to a spring where fairies meet, and he receives gifts from them. If it is accepted that the mother-deities of paganism, represented by the fairies, have been replaced by the Virgin of Christianity, it is not surprising to learn that the Chartreuse was dedicated to the Virgin-Fairy, as if it were necessary to commemorate the old deity of the forest.

■ THE FOREST WORLD OF THE OUTLAW

The word 'forest' is derived from the Latin *silva forestis* or 'forest outside the enclosure'. It is an essentially strange place to the outsider; a place where hermits and outlaws live. Robin Hood, the legendary Saxon hero (based on a historical person) was banished by the Normans and, with Maid Marian, was forced to live in Sherwood Forest. The forest is a place where the laws of society no longer apply because it comes under another authority.

Other similar stories about outlaws exist. Tristan and Yseult were forced to live in the sinister forest of Morrois. Condemned to death by King Marc, the lovers were obliged to hide and to forage for food. Béroul, the Anglo–Norman poet, never presents the forest as a paradise or a place of innocence, but as a place of suffering. Life in the forest condemns a person to solitude, abandonment, and poverty. In the Middle Ages, life in the forest was a prelude to death. To shun society was condemned, for society in the Middle Ages was profoundly opposed to individualism: people could not exist unless they belonged to a group, a brotherhood, a parish. It was the community which gave the individual a reason to exist, and which defined his social status.

It would however be facile to claim that the 'wild man' simply represents the 'other', the 'other place' or the 'other time'. He is not a simple representation of 'otherness', but represents the ambivalence of untamed nature, beyond what is good or what is evil. He embodies the animal and the outlaw in man, or the human aspect of the animal forest. The wild man of the woods is not only (because of his repulsive or deformed appearance) a cursed, marginal and solitary being, but is at the same time a figure of wonderment because of his mysterious gifts. The wild man is a mediator: it is he who shows the knight the way to the fountain, and it is he who reveals to humans the secret path to the world beyond, the world of the supernatural. The medieval forest can be a place of enchantment because it is also a place which defies the laws of the ordinary world; a place where all is possible, be it good or evil.

The medieval forest, hostile and menacing in its elemental force, the 'perilous forest' as it is often called in the Romance texts, can easily devour its human prey. In folklore, the dwelling of the ogre is always to be found in the very depths of the forest. In the story of 'Hop o' my Thumb', the woodcutter's children are led into the thickest and darkest part of the forest and are surrounded by the howling of wolves coming to devour them. In the deepest part of the forest they come upon the abode of the ogre, the master of this dark and frightening place. In the Romance of Tristan, the forest of Morrois is inhabited by a hermit named Ogrin, who, in spite of his gentle appearance, conjures up the old figure of the ogre. Indeed, the forest is the kingdom of the ogre, as is shown in many European and Japanese folk tales studied by Shinoda (1994). It is here too that the werewolves prowl, as related by the poet Marie de France (Bisclavret, v. 9-12):

Garvalf, ceo est beste salvag
Tant cum il est en cele rage
Humme devure, grant mal fait
Es granz forez converse et vai.

Werewolf is a savage beast
In this bestial state
He will devour men and do evil
He roams in the great forest.

THE FOREST OF MERLIN

The character of Merlin is certainly the most famous man of the woods in the literature of the Middle Ages. He can be seen as a transitional character between the fairy and the sorcerer. The forest becomes his natural habitat. Victim of a curse, he is condemned to live in the Caledonian forest like a wild animal. He sometimes uses a stag as a mount and in the summer he lives in the company of a grey wolf.

In 'Vita Merlini', a twelfth century Latin text, Merlin lives in the depths of the woods with his sister Ganieda. His dwelling place has seventy doors and seventy windows through which the sky can be seen (Clarke, 1973). This extraordinary astronomical observatory is used for a totally different world of knowledge. Merlin is shown as the possessor of knowledge beyond the understanding of ordinary mortals. It is significant that this knowledge is relegated to the fringes of the wilderness world in which Merlin lives, for in the Middle Ages science is often linked to magic and the magician is a marginalised being, set apart from the community. A place apart, the forest harbours all the potential for widening knowledge, and there the forces of magic retain their powers.

In the legendary forest of Brocéliande there are a number of megalithic monuments which, according to legend, indicate the preference shown for the forest by magical beings, and Merlin in particular (Bellamy, 1895). It is in the heart of this forest under the remains of a megalith, that the body of Merlin is said to lie. It is there that the Fairy Viviane is said to have tricked Merlin and trapped him, taking advantage of the passionate love that he felt for her. After leading him to the edge of a hole in which he lay down, Viviane caused two large stones to fall on him. Merlin thus became the prisoner of the fairy with whom he had fallen in love. Another version of the story relates that Viviane enclosed Merlin in an invisible prison of air. She then prepared his tombstone nearby, which is believed to be a megalith known as the 'Hotié de Viviane'. The monument dates to approximately 2500 BC and is elliptical in shape, being constructed of leaning stones supporting one another. Flints, pottery sherds and small beads of pierced pebbles have been found there.

THE MAGIC OF WITCHES

Sorcerers and witches are the descendants and heirs of the ancient deities of the forest. This aspect is often neglected in the study of witchcraft because the folklore of witches was later distorted by the inquisition and the repression to which they were subjected. There are myths about witches (not to be confused with the mythology of witchcraft) which have led to a misunderstanding of the phenomenon since the Middle Ages (Eliade, 1976). For example, the word 'sabbath' causes confusion both because of its Biblical meaning and its meaning in witchcraft (Ginzburg, 1992).

The rites of witchcraft provide evidence of contact with the magical forces of the forest and of nature in general. Witches try to maintain a magical consciousness in which the link between man and nature is intact. Some historians have categorised witchcraft as an expression of fertility cults (Delumeau, 1978: 474). Ginzburg (1966) refers to the tale of the good witches of Frioul in Italy. Armed with fennel stalks, they journey in their dreams or in a semi-sleep to the Valley of the Sorcerers. With their fennel in their hand, they take up battle against the evil witches who brandish stalks of sorghum to defend themselves. If the good witches are victorious, the harvest of wine and wheat will be abundant; if they fail, famine will ravage the land. Ginzburg links these traditions to an ancient cult of the dead and to a belief in the journey of souls illustrated by 'The Wild Hunt'.

Generally, witches know the secrets of herbs

and plants. Like Medea, they exploit their magic qualities, making potions for different effects. This skill was shared by the fairies. The mother of Yseult made a love potion from herbs for King Marc and Yseult, and the fairy Morgane healed wounds with the aid of herbs and plants. The traditional pharmacopoeia used many remedies based on plants, and was at one time the only known medicine. Today, ethno-pharmacological laboratories, such as the European Ecological Institute of Metz, study the old medieval remedies based on medicinal plants and sometimes discover hidden, and so far unexplained, virtues in these traditional cures.

From the High Middle Ages (Walter, 1989: 270) to the eighteenth century (Delumeau, 1978: 484) and even beyond, the customs involving the herbs of Saint John persisted in spite of ecclesiastical condemnation. There is ample testimony to show that forest plants possess unsuspected qualities from which man can benefit. From her ancient pact with the trees and the plants, the witch held these secrets, inaccessible to ordinary mortals who failed to maintain the vital link with the natural wilderness.

A good example of the healing virtues of the forest sanctuary in the Middle Ages is seen in the cult of a mysterious saint of the thirteenth century who was said to be a dog in his original incarnation. An Inquisitor of the thirteenth century condemned the rite practised by women in a forest in the east of France who placed their sick, or sometimes stillborn, children on the tomb of the holy dog, and would then invoke the fauns of the forest to restore health to the small bodies presented to them. The report of the Inquisitor reads:

> It was particularly the women with sick or retarded children who went there. They carried them in their arms and first went to a castle, about a league away, where they met with an old witch who initiated them, teaching them the rites they should follow and the manner in which they should invoke and offer up sacrifices to the evil spirits. The witch would accompany the mother and when they reached the tomb she would offer up salt and diverse other things. Then they would hang the baby's swaddling clothes on the surrounding bushes and stick needles in the trunks of the trees, which had grown around the tomb of Saint Guignefort. That done, the mother would stand on one side of a gap between two trees growing close together, and the witch would stand on the other side. Then they would throw the naked child from one to the other nine times, so that the baby passed through the space between the trees. Doing this they would invoke the evil spirits and ask the fauns of the forest to take the child, sick or retarded, and return it full of life, plump and healthy [Saintyves, 1987: 820]

Sometimes the child did not respond to this treatment and the Inquisitor deplored the many accidents that occured which often resulted in the death of the child. For this reason he decided to exhume the remains of the holy dog, and to cut down the sacred wood and burn the bones. Then, with the threat of punishment, he forbade Christians to come to the place and to indulge in these superstitious rituals. This testimony of the Inquisitor, Etienne de Bourbon, is most informative about the Church's repression of ancient pagan beliefs linked to the world of the forest. The Church of the thirteenth century tolerated less and less the belief in fauns and other deities and the suspect mediation which the witches provided. The obscure powers of the forest worried the Church which thus sought to eliminate them.

Another historical example demonstrates the growing fear of the forest world at the end of the Middle Ages, when towns were becoming more important. During her trial, Joan of Arc was interrogated on the exact nature of her relations with the fairies (Duby and Duby, 1973: 41–2). Like all the girls of her village, she went to the place in the forest where the 'tree of the ladies' (another name for the 'fairies' tree') grew. Near this tree was a spring and those who were sick with fever used to drink the waters, hoping to be cured. She had heard it told by the old folk that the fairies lived there. Joan was also questioned about the Hoary Wood—'the wood of the oak trees'—mentioned in the Prophecies of Merlin. Certain of these prophecies which Joan knew, said that out of the Hoary Wood would come a maiden who would work miracles. Various witnesses depicted her as this providential maiden who had emerged from the Hoary Wood like a witch, to cast a spell on the English in order to chase them out of France and put an end to the Hundred Years' War. There is no doubt that the judges who condemned her to the stake claimed that she had the soul of a witch because she had maintained secret relations with fairies who bewitched those with whom they came in contact.

Thus, in certain rural districts as late as the fifteenth century, the forest still represented a place of hope and healing. The trees and the springs were a calming presence and the forest had a healing power. However, at the same time, the urban élite became more and more suspicious of the forest as it had become a place where rabies reigned. Hunting mythology (Hell, 1994) suggests the idea of 'black blood', the blood that is in the veins of the wild men of the woods and in those of the werewolf or the possessed spirit. To hunt was to risk contamination by this 'black blood' and being devoted to hunting was to surrender to the 'wild flow' (*le flux sauvage*) and risk falling under its influence. Hunting was perilous because of the ever-present danger of the forest, which housed this 'black blood'. There was also the danger of madness, as is suggested in many forest myths.

The forest was also a favoured place for sabbaths, although not the only one. As Gaignebet (1993: 50) reminds us, it is in the middle of the night of Walpurgis, the night of the first of May, that the witches take to flight:

> This is the night when throughout all Europe the broad-leaved trees take over. But the colour of the leaves changes also at Saint John's-Tide in the summer and again, on the first of November. Pliny explains to us that it is at the solstices that the leaves of the two-coloured trees, the poplar, the willow and the olive, will turn. Those that appear light-coloured the day before are dark the following day. During these nights when the leaves change colour, you have only to pronounce a short spell just a fraction too early or a fraction too late, to find yourself buried under the leaves. For it is during these nights, according to Saint Gregory the Great (v. Dialogues), that the souls of the dead rest on the leaves. When we clothe ourselves in leaves, as they turn, they communicate to us the sighs of the dead. That is what we learn on the first of May, that is what the weather teaches us when the first leaves fall on the eleventh of November and the sabbath of Saint-Martin.

By associating the tradition of the broad-leaved trees, the wild men of the woods disguised by leaves, with the sabbath of the witches, Gaignebet has stressed the close relationship that links these two rites with the theme of the journey of departed souls.

The Natural Religion of the Witches

Witchcraft is usually associated with the Middle Ages, but although it first appeared at this time, it actually reached its zenith in the sixteenth and seventeenth centuries. By 1868, the historian Jules Michelet realised that witchcraft had not grown out of nothing. To him its origin was to be found in a system of pre-Christian beliefs and naturalistic religious practices which the Church had not succeeded in Christianising, and which it therefore sought to combat. Christianity, in becoming a scholarly and intellectual religion, had broken the link with the cosmic and natural world that was the source and strength of pre-Christian and shamanist religions.

This new, abstract, Christian religion, practiced in the towns and villages, discouraged the earlier rural Christianity which it had at first encouraged: a rural Christianity which had set its chapels in the woods, near the springs or the fountains of the fairy deities. This discouragement led to a reaction in support of the followers of the ancient divine forces. The witches sustained their intimacy with the elements of nature: fire, air, water, and earth, thereby retaining powers that other men had lost, apart from the priests who now placed their special powers under Church authority.

Witches could ward off tempests (Paravy, 1982: 67–71); the elements obeyed them. The witchcraft trials were the result of this belief. Every time economic or social misfortune struck, it was attributed to the evil influence of witches. By using the forest resources, witches were believed to develop unexpected uses for everyday objects such as the broom. Believed to be used for taking them to the witching hour, the witches' magic broomstick was made from slender branches and twigs of trees, either the birch, the most favoured shamanist tree, or the broom (Brosse, 1989). Between 1440 and 1442, Martin Le Franc composed a long allegorical poem entitled *Le Champion des Dames* in which this belief is described (Deschaux, 1983). A great debate between the protectors of women and their adversaries makes reference to a disturbing scene in which the witches clearly plead against the cause of women:

Vrai est, ouy l'ai je, m'encrois
Que les vieilles, ne deux, ne troi
Ne vingt, mais plus de trois milliers
Vont ensemble en aucuns détroits
Voir leurs diables familiers
Je te dis avoir vu en chartre
Vieille laquelle confessait
Aprés qu'écrit était en chartre
Comment, des le temps qu'elle était
De seize ans ou peu s'en fallait
Certaines nuits de la Valpute
Sur un bâtonnet s'en allait
Voir la synagogue pute
Dix mille vieilles en un fouch
Y avait-il communément
En forme de chat ou de bouc
Voyant le diable proprement
Auquel baisaient franchement
Le cul en signe d'obéissance.

It is true, I have heard it and believe it
That old women, not two, nor three
Nor twenty, but more than three thousand,
Go together to a special place,
To meet with their familiar evil spirits.
I tell you I saw in prison
One old woman who confessed
Her testimony was written down
How, from the time that she was
Sixteen or less, she went
On nights of St Walpurgis
Travelling on a broom
To see a monstrous sabbath.
Ten thousand old women at a crossroads
There were in the congregation.
In the shape of a cat or a goat
They saw the devil himself
And they willingly kissed
His arse as a sign of obeisance.

According to this text, witches form a secret society meeting together in wild remote places, far from the eyes of the world. Good people are frightened by their sheer numbers, as much as by their practices: the flight through the air on a broomstick, their tryst with the devil disguised as a cat or goat, and the kiss of homage to the Master of the Sabbath. There has been much discussion about the name Valpute in the text. According to Hansen (van Gennep, 1991: 83), it is a valley in the Dauphiny where a great witchcraft trial took place in 1437. However, as has been clearly demonstrated by van Gennep (1991: 84), Valpute is really a variant of Walpurgis. There is a clear link with the well known feast of Walpurgis Night, sometimes called the Night of Diana or the Night of Herodiades. During this night of 30 April–1 May, thousands of witches congregated at the summit of the Broken, a mountain in the Harz range (van Gennep, 1991: 84). However, the word Valpute can also be read as the 'Valley of the Whores' (*vallée pute*) which does not identify a particular place, but was rather a place of perdition where witches, the symbols of infamy and perversion, congregated. According to their detractors, they left traces of their filthy practices at all their meeting places.

The forest is sometimes destroyed by the work of the witches. The well-known symphonic poem *A Night on the Bare Mountain* by Moussorgski was inspired by a Russian folk tale, and refers both to a ritual space and time for witches. The space is the Bare Mountain, a cursed and sterile desert, marked by relentless doom. The time is the Night of Saint John, well known for its strange and magical associations. The Summer Solstice is observed as a special time when hidden cosmic forces affect the movement of the planet like a Solstice clock, regulating the Seasons.

An anonymous lay of the thirteenth century refers to a procession of horses in the sky at the time of the Feast of St John. Although this event may have had more to do with the world of fairies than with the world of witches, these two opposites represent the two sides of the same reality. The lady friend of a knight falls asleep in an orchard after saying her prayers. During her sleep, she travels through the sky and wakes up suddenly beside her knight, near a dangerous ford, unable to remember the slightest detail of her aerial flight (Walter, 1989: 489–92). This theme is common in folk tales, and is particularly connected to the twelve days between Christmas and Epiphany: 'One avoided going into the woods and the forests after sundown for during these nights the spirits were abroad' (de Westphalen, 1934: 495). These roaming spirits were said to practise clairvoyance.

There can be no doubt that this refers to an old shamanist practice, even if the exponents of the practices were not always conscious of this. Sometimes considered as pure hallucination, these nocturnal horse-rides represent the survival of the ancient belief in fairies, which had been demonised into witches. They go back to the old myth of the journey of the soul, which can, in certain conditions, leave the body and return to it.

Finally, it can be asserted that witchcraft is an invention of Christianity which sought to demonise and thus exterminate the remaining traces of the old calendar of pagan rituals. Witch mythology is thus merely one aspect of pre-Christian mythology and does not constitute a separate tradition. The rites and myths of witchcraft are part of a vast collection of pre-Christian myths, traces of which can be found in medieval hagiography, but which more often find their points of comparison in the great shamanist traditions.

In conclusion, the world of the forest is a rich minefield for the study of mythology in the Middle Ages. It reflects Medieval thought which sought to establish dialogue with the obscure forces of the forest and was captivated by its relationship with the spirit world. However it also expresses a secret concern and anxiety about this other world that threatened the intellectual certainties which Christianity was trying to consolidate. Christianity increasingly cut its links with the sacredness of the natural world in order to develop as an abstract religion of inner values. The great upsurge of rationalism in the seventeenth and eighteenth centuries served only to hasten this despiritualisation of the forest, a movement which Christianity initiated and promoted. Belief in the sacredness of the forest was condemned. The forest became an economic space, colonised by society. The spirit of the forest was rejected by society in all its major institutions, the city, the town, the family, and religion.

In the eighteenth century, in his *Philosophical Dictionary* under the heading 'he-goat', Voltaire wrote triumphantly, 'We have already said that more than one hundred thousand alleged witches have been put to death in Europe. Philosophy itself has at last cured man of this abominable chimera and has taught judges that one should not burn imbeciles.' In fact, if fairies and witches began to disappear at the beginning of the eighteenth century, it was less the effect of the Enlightenment rationalist ideology combating the so-called obscurantism of the Middle Ages, than the effect of the extension of the road network and the systematic exploitation of the forests. In becoming an economic space, the forest lost its supernatural soul. The systematic cutting of wood for heating and the process of deforestation gradually removed from human consciousness the legendary beings and fairies who peopled the forests and the ancestral beliefs that surrounded them.

In the sixteenth century, the poet Ronsard (Ronsard, Elegies, XXIV) cried out to the woodcutters of the forest of Gastine who were destroying his cherished wood which had been sold by the King of Navarre (the future King Henri IV):

Écoute, bûcheron, arrête un peu le bras!
Ce ne sont pas des bois que tu jettes á bas
Ne vois-tu pas le sang, lequel dégoutte à force
Des nymphes qui vivaient dessous la dure écorce?
Sacrilège meurtrier, si on pend un voleur
Pour piller un butin de bien peu de valeur
Combien de feux, de fers, de morts et de détresses
Mérites-tu, méchant, pour tuer nos déesses?

Listen, woodcutter! Stop your arm awhile
These are not woods which you are felling
Do you not see the gushing blood
Of the nymphs who dwelt under the strong bark?
Ungodly murderer, if we hang a thief
For stealing booty of very little value,
How many fires, irons, deaths and sufferings
Do you deserve, evil one, for killing our goddesses?

Is not Ronsard here explaining to us that in destroying the forests, we are also destroying the ancestral myths of our civilisation and condemning ourselves to becoming rootless people? What if the forests were one day to take their revenge?

References

Bellamy, F. (1895): *La forêt de Bréchéliant, la fontaine de barenton, quelques lieux d'alentour, les principaux personnages qui s'y rapportent*, Rennes, Plihon & Hervé.

Bernheimer, R. (1952): *Wild Men in the Middle Ages: A Study in Art, Sentiment and Demonology,* Harvard University Press, Cambridge.

Bligny, B. (1984): *Saint Bruno, le premier chartreux,* Ouest-France, Rennes.

Brosse, J. (1989): *Mythologie des Arbres*, Plon, Paris.

Chateaubriand, F. R. de. (1989): Mémoires d'Outre-Tombe (1840), Garnier, Paris.

Clarke, B. (1973): *Life of Merlin by Geoffrey of Monmouth*, edited with introduction, facing translation, textual commentary and name note index by Basil Clarke, Cardiff, University of Wales Press for the Language and Literature Committee of the Board of Celtic Studies.

Cohn, N. (1982): *Démonolâtrie et sorcellerie au Moyen Age: fantasmes et réalitès* (trans. from English by S. Laroche and M. Angeno), Payot, Paris.

Delumeau, J. (1978): *La peur en Occident (XIVe-XVIIIes.)*, Fayard, Paris.

Deschaux, R. (1983): Oui ou non, les sorciéres volent-elles? *Recherches et Travaux (Grenoble)*, 24: 5–12.

De Westphalen, R. (1934): *Dictionnaire des Traditions Populaires Messines*, Le lorrain, Metz.

Duby, G. and A. Duby, (1973): *Les procès de Jeanne d'Arc,* Gallimard-Julliard, Paris.

Duval, P. M. (1993): *Les dieux de la Gaule,* Payot, Paris.

Eliade, M. (1949): *Traité d'histoire des religions*, Payot, Paris.

—— (1963): *Aspects du Mythe*, Gallimard, Paris.

—— (1965): *Le sacré et le profane*, Gallimard, Paris.

—— (1969): *Le mythe de l'éternel retour,* Gallimard, Paris.

—— (1976): *Occultism, Witchcraft and Cultural Fashions*, University of Chicago, Chicago.

Evola, J. (1976): *Métaphysique du sexe* (French trans.), Payot, Paris.

Foulon, C. (1949–52): Enchanted Forests in Arthurian Romance, *Yorkshire Celtic Studies,* 5: 13–18.

Gaignebet, C. (1974): *Le Carnaval*, Payot, Paris.

—— (1993): Discours de la sorcière de Saint-Julien-de-Lampon. In N. Jacques-Chaquin, (ed.), *Le Sabbat des Sorciers (Xve-XVIIIes.),* Millon, Grenoble, pp. 47–54.

Ginzburg, C. (1966): *I Benandanti: Stregoneria e Culti Agrari tra Cinquecento e Seicento*, Einaudi, Turin.

—— (1992): *Le sabbat des sorcières* (French translation), Gallimard, Paris.

Guyonvarc'h, C. (1986): *Les Druides,* Ouest-France, Rennes.

—— (1990): *Nemeton: la forêt sanctuaire in Brocéliande ou l'obscur des forêts,* La Gacilly, Artus, pp. 34–6.

Harf-Lancner, L. (1984): *Les fées au Moyen Age. Morgane et Mélusine*, Champion, Paris.

Harrison, R. (1992): *Forêts. Essai sur l'imaginaire occidental* (French translation), Flammarion, Paris.

Hell, B. (1994): *Le sang noir: Chasse et mythe du Sauvage en Europe,* Flammarion, Paris.

Lecouteux, C. (1995): *Démons et génies du terroir au Moyen Age,* Imago, Paris.

Le Goff, J. (1985): *Le désert-forêt dans l'Occident médiéval: L'imaginaire médiéval*, Gallimard, Paris, pp. 59–73.

Meyer, P. (ed.) (1875): Brun de la Montagne, *Société des Anciens Textes Français,* Paris.

Micha, A. (1992): *Lais féeriques des XIIe et XIII siècles,* Flammarion, Paris.

Michelet, J. (1868)*: La sorcière*, ed. P. Viallaneix (1966), Garnier–Flammarion, Paris.

Muchembled, R. (1979): *La sorcière au village (XVe-XVIIIes),* Gallimard–Julliard, Paris.

O'Hara Tobin, P. M. (1976): *Les lais anonymes des XIIe et XIIIe siécles,* Droz, Genève.

Paravy, P. (1982): Prière d'une sorcière du Grésivaudan pour conjurer la tempête (procès d'Avalon), *Le monde alpin et rhodanien*, Grenoble, Centre alpin et rhodanien d'ethnologie, pp. 67–71.

Printz, A. (1956): *Hombourg sur Canner,* Le Lorrain, Metz, pp. 71–72.

Roheim, G. (1974): *La panique des dieux* (French translation), Payot, Paris.

Saintyves, P. (1930): *En marge de la légende dorée: songes, miracles, survivances* (1987 rpt), Laffont, Paris.

Shinoda, C. (1994): *La métamorphose des fées: Etude comparative des contes populaires Français et Japonais (autour des contes sur le mariage entre humains et non-humains*), Nagoya.

Van Gennep, A. (1991): *Les Hautes Alpes traditionnelles*, Curandera, Voreppe.

Walter, P. (1989): *La mémoire du temps: Fêtes et calendriers de Chrétien de Troyes à La Mort Artu*, Champion, Paris.

—— (1992): *Mythologie Chrétienne: rites et mythes du Moyen Age,* Editions Entente, Paris.

Deciduous oak with mistletoe (Wien, Austria. Photo by Yasuda)

CHAPTER 11

The Insect and the Western Image of the Forest[1]

ANDRÉ SIGANOS

Introduction

As a starting point, I will demonstrate the archetypal image of the insect: the sum of the symbolic characteristics that have been attributed to insects by numerous cultures since remote antiquity. This will enable a better understanding of the close links between insects and trees in an analysis of three well-known nineteenth-century stories. I will then expand and broaden the symbolic function of the insect and the tree by reference to the essential and archaic image of the labyrinth, which has inspired many contemporary narratives, including Japanese ones. In conclusion, I will consider the distinctive feature of the insect, the tree, and the forest: that of initiatory metamorphosis.

The Archetypal Image of the Insect

From a phylogenetic point of view, the insect is a radical and distinctive phenomenon. It is highly adaptable and constitutes an ancient biological model that has developed through a different evolutionary process than that which engendered *Homo sapiens*. The insect first appears in the fossil record during the Palaeozoic. Not only the first to evolve, insects also surpass all other animal species in their variety and number,[2] in their diversity of habitats, and in their capacity to survive extreme conditions. For example, a tic can survive for nine years without food, certain *Coleoptera* are capable of developing in seventy-degree alcohol, and the larva of *Diptera* can reproduce in crude oil. Insect statistics are astonishing: a queen termite can lay up to forty thousand eggs in a day, producing in one year living matter equivalent to the weight of a fishing trawler![3] Such feats, coupled with the hierarchical and complex societies of some insect species, explains why these creatures appear in philosophical, spiritual, and psychoanalytical domains.

THE *BLACK* INSECT

'They are the things that rattle doors on nights of wind, the faces that watch half-seen from the deepest woods.' This quotation from Keith Roberts (1975: 220) in his novel about giant wasps, encapsulates how profoundly mysterious and inexplicable this, and other, insect species have been to humans.

Humans project on to insects an image that can be best summarised as *black*. They are commonly associated with death—as in Vietnam and France, for example, when the sound of the Cicada is heard (as Saint-John Perse, a French poet, wrote: 'hear death come alive with the chirp of the cicada' (*Amers, Strophe:* 1972: 336). In addition, the insect is frequently perceived as hideous: it can live and mate when part of its body is amputated; it can feign death for long periods; and it can mimic a branch or a flower. Migrating insects (particularly locusts but also, surprisingly, certain butterflies) sometimes swarm in such large numbers that they cause considerable environmental damage. In Texas in 1921, a swarm of *Lepidoptera* was 400 km wide, weighed about 60,000 tons, and contained approximately thirty billion butterflies (Lhoste, 1979: 69).

American cinema enthusiastically took hold of this theme: a giant spider featured in *Tarantula* (1952), giant ants in *Them!* (1953), and *Phase IV* (1974); there have been numerous other insect-related films such as *Bug* (1963), *The Swarm* (1972), *The Giant Spider Invasion* (1978), *Alien* (1979), and *The Fly* (1986). In such stories, the strength, intelligence and aggression of the insect intrigues us. The colour of the insects is significant and is usually, for example, black and gray for the flies that invade the earth, black and yellow or black and orange for the *Hymenoptera,* that suddenly emerge from deep crevasses, a deep and shiny black for fireflies, and a mat and sinister black for the radioactive or gigantic spiders. The image of the black insect in science fiction, fantasy and horror stories (see Siganos, 1985: 91–109) associates the insect with terrible forces: often those that emerge from darkness. In these stories, people are impelled to become satanic in order to fight the insects, whose attack is frequently foreshadowed by the noise they make.

A different set of images is due to the fact that, apart from humans, insects are essentially the only other social animals. This parallel has prompted the notion that insect societies may be disturbing precursors of future human society. Leaving aside the frequent comparison that has been made between the city and the anthill, perhaps there is also a relationship between high population density and the accompanying change in human behaviour (such as heightened aggression and capacity for organisation). High concentrations of insects of the same species causes them to modify their colour, growth-rate, and even their fecundity (Chauvin, 1961: 32).

If one turns to the nuptial habits of the insect, the prospect is scarcely more edifying, since death seems to frequently accompany mating: the drone bee leaves his sexual organs in the abdomen of the queen bee; the male praying mantis, and sometimes the male spider, is eaten by the female of the species. The dung beetle dies after digging a pit for the female to lay her eggs in. The significance to humans of the dominating and castrating female insect, highlighted by the social behaviour of these creatures, appears to be confirmed by psychoanalysis. For example, the appearance of the spider in a dream can be 'the symbol of the bewitching female, of the satanic virago, whose overriding aim is the destruction of the male' (Aeppli, 1978: 286). The behaviour of the female insect has thus been associated with the castration complex: a complex found in myths, stories (Shinoda, 1994), and even in surrealist painting (Dali depicts putrefying kisses in association with locusts, flies and ants in the paintings he created in 1929).

■ THE INSECT AND THE TREE

In many parts of the world, honey is seen as a primary source of food, and the bee is often deified. Hesiode (1928: 232–3) explains that at the dawn of creation 'the oak bore acorns at the top, with bees in the middle'. Insects appear in numerous historical documents, such as the Veda, the Bible, and the Koran, and frequently figure in Hittite, Ancient Egyptian, and Christian mythology, as well as in that of the Indians of South America (Siganos, 1985: 21–24). In a Kayan myth from Borneo, a spider is at the origin of the first tree and then of the first man (Poignant, 1968: 72). An Apinayan myth asserts that, during the Flood, men took refuge in the trees and became insects (Lévi-Strauss, 1964: 387).

A close symbolic relationship appears to be frequently fostered between the tree, as the axis of the world, the insect, as the producer of an image of this axis (the spider's web), and the image of the 'Omphalos', the centre of the world. The termite nests among the Bantu or among the Bambara (Heusch, 1972: 24; Zahan, 1965: 175–87), the Ananse spider among the Fanti, or the oceanic myth of Tawhaki, all demonstrate the extent to which the spider and its web bring out the ascensional symbolism and the notion of *axis mundi* (Durand, 1960: 138; Calame-Griaule, 1969–70: 2, 120).

The Delphic Pythia, an important cultural centre in Ancient Greece, was known as the 'bee', the priests of Ephesus, another cultural centre of the first rank, were called 'king-bees' (Steiner et. al., 1958: 72; Pindar, 1922: 60; Pausanius, 1965: 13; Frazer 1966: 135–36). The Druidesses who cut mistletoe in oak trees were known as 'bees' (Pliny the Elder, 1952: 249; Chevalier, 1969: 384) and Rebecca's nurse, 'Deborah the Bee', was buried under the 'Weeping Oak' (Genesis, 35, 8). There was, therefore, an essential relationship between the divine, the bee, and the axis of the world. For some pre-literate and (at least up until the end of the seventeenth century) Western peoples, the bee was a messenger of the gods, a creature testimony to divine wisdom (Marchenay, 1979: 129).

The Insect and the Forest in the 'Learned' Tale

This 'archetypal' image of the insect elucidates the way in which this image is interpreted in the following three 'learned' tales,[4] and how this image has changed and perpetuated itself in Western imagination.

■ THE CHILD STRANGER

In *The Child Stranger* by Hoffmann (1832) the character of Master Ink is endowed with a negative symbolism. On the other hand, the characters of Thaddeus Von Brakel, his wife, and two children (Felix and Christlieb)[5] represent 'Good', and live frugally but are nonetheless happy. The two children are a model of happy childhood. A visit by their uncle, Count Cyprien, a minister at court, accompanied by his wife and two children, allows the author to portray a diametrically opposite picture. The Count's children are well educated but filled with the sort of knowledge that is only found in books, and their good manners do not prevent them from being hard or cruel. Count Cyprien is thin and sinister, affected and vain, precisely the opposite of his cousin who is a simple country gentleman, and is natural, cheerful, and modest. This system of opposites constructed by the author even applies to the aggressive toys which the rich children offer their poor cousins. The apparently splendid and sophisticated puppets which the children receive merely mimic life (as do the gift-givers), rather than living it in the fullest sense.

Soon, the whole family feels that the visit has introduced unhappiness. The unhappiness is however ambiguous: it constitutes the first awareness of a foreign world; the children become aware of their ignorance and lack of culture. A second initiatory process occurs when a child ('Child Stranger') introduces himself. The emotion they feel towards the intruder can not be easily categorised as either good or evil (e.g. Hoffmann, 1832: 121).

The appearance of the Child Stranger in a dream is an inexhaustible source of pleasure for the children because it helps them to understand why nature is such a powerful force of attraction. The Child Stranger however also shows them that there is no happiness without danger because he himself is haunted by a big black fly, 'the sombre king of the gnomes'.

From this point in the story, the reader begins to understand the behaviour of Master Ink, Count Cyprien's envoy, who pricks Thaddeus' children with a needle hidden in his hand on their first encounter. By the skinniness of his legs and his big black belly, this envoy is as much a representative of the spider as of the bumblebee. He clumsily flies away when he metamorphoses into a bumblebee, after having asked the children how far they have progressed in the world of knowledge. United by an infernal pact, Cyprien and the 'king of the gnomes' later manifest themselves through other intermediaries, in particular the toys which Felix and Christlieb broke in the wood, and which grotesquely come to life and threaten them.

If we try to interpret this tale, what do we find? From a moral perspective, the combat which takes place between the king of the gnomes and the child-fairy, represents the struggle between Good and Evil in the children's conscience, in which dream and reality are so easily intertwined. However the tale is also marked by an ambiguity, which calls into question the manichean opposition between nature as the source of virtue, and a human world ridden with vice. In this respect the satanic aspect of Master Ink is an essential element in the story.

The forest, so prized by the children and so hated by Master Ink, plays a special initiatory role and is also an illusory place of refuge. Each time the children come out of the forest, they have to relive the problems they encountered there, including the basic question of the relationship between dreams and reality. If the insect represents evil, we need to look more closely at the true nature of this blackness. Is it a form of basic evil which the reader can openly identify as such or is it a representation of the aggression which every man and child carries within themselves? It is not a matter of chance that the Child Stranger is androgynous and thus conforms to the aspirations of both the little boy and the girl. The bumblebee, on the other hand, personifies the harsh reality of work with the attendant moral lesson that 'you get nothing if you do nothing to earn it'.

■ L'HISTOIRE DU VÉRITABLE GRIBOUILLE

With the *Histoire du Véritable Gribouille*, by George Sand (1850), the complementary symbolic role of the insect, the tree, and the forest is reinforced and enriched. The story itself is partly based on *The Child Stranger* (see Siganos, 1982), as Sand greatly admired Hoffmann. Gribouille is one of seven children belonging to Bredouille and Brigoule, parents who are as mean as they are unscrupulous. In spite of the bad treatment he has endured at home, Gribouille refuses to become a liar and a thief. One day, while resting at a crossroad under the shade of a great oak tree, the child is stung by a bumblebee but does not kill it, trying instead to explain the nastiness of the act that the bee has perpetrated. When he wakes up, Gribouille finds himself confronted with Monsieur Bourdon ('Mister Bumblebee') who has come to thank him for his good act and to give him 'some spirit' (Fig. 1). After transforming himself into an insect, Bourdon disappears.

When he returns home, Gribouille learns that Bourdon was really intending to make his fortune and his parents promptly order him to go back and find him. Gribouille sets out, but, feeling tired, he stops under a fig tree and notices several honeycombs in its branches. When at last he arrives at the castle of Monsieur Bourdon where a large party is taking place, he soon falls asleep, or at least he thinks he does, for in a state of trance he witnesses the transformation of a valet into a giant spider who proceeds to ensnare him in his web (Fig. 2). Against his will, Gribouille becomes the adopted son of Monsieur Bourdon who initiates him into his secrets. In order to do this he takes him to the crossroads where the oak tree is found. There the child eats some of the tree's acorns, which put him to sleep, and he begins to dream about a furious combat between the people of the honeybee and the people of the bumblebee. When he wakes up he refuses to become an evil spirit, as Monsieur Bourdon asks, before once again changing into an insect and chasing the child (Fig. 3). Gribouille jumps into a stream and is transported down to the sea where he changes into an oak branch (Fig. 4). It is in this

Fig. 1: Gribouille and Monsieur Bourdon (drawing by Maurice Sand in Sand, 1957)

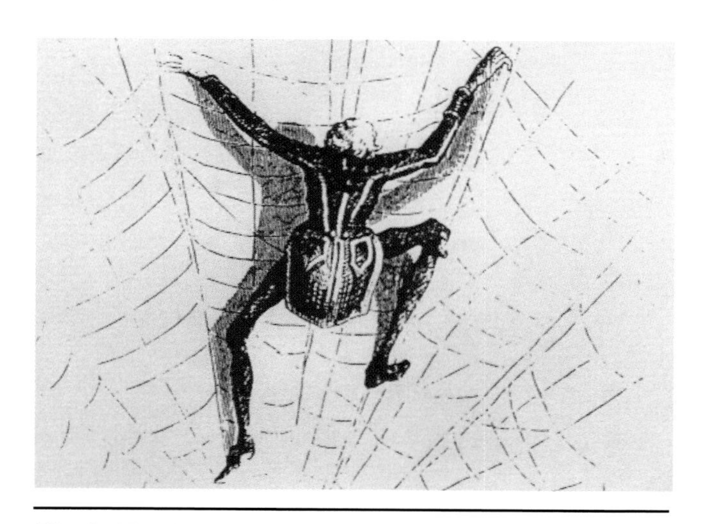

Fig. 2: The Spider Valet (drawing by Maurice Sand in Sand, 1957)

transformed state that he is washed up on the island of good spirits.

The subtitles to the first and second parts of the story ('How Gribouille threw himself into the river out of fear of getting wet' and 'How Gribouille threw himself into the fire out of fear of being burned') give considerable prominence to the two initiation trials that the child must undergo in order to reach a certain level of wisdom. There is no way of avoiding these trials and, in effect, they constitute an essential part of the formative process, which every human must undergo. Importance can also be attributed to the child's periods of sleep and the metamorphoses he witnesses or experiences. In the first part, Gribouille strives hard to win the love of his parents and, in doing so, passes constantly between a state of dreaming and one of conscious reflection; it is in an intermediary zone that the child discovers the correct attitude. The sting of the bumblebee appears to play a role in the first initiation: not to pain, but to the virtue of self-control in the face of unjustified external aggression, and, above all, as an initiation into the

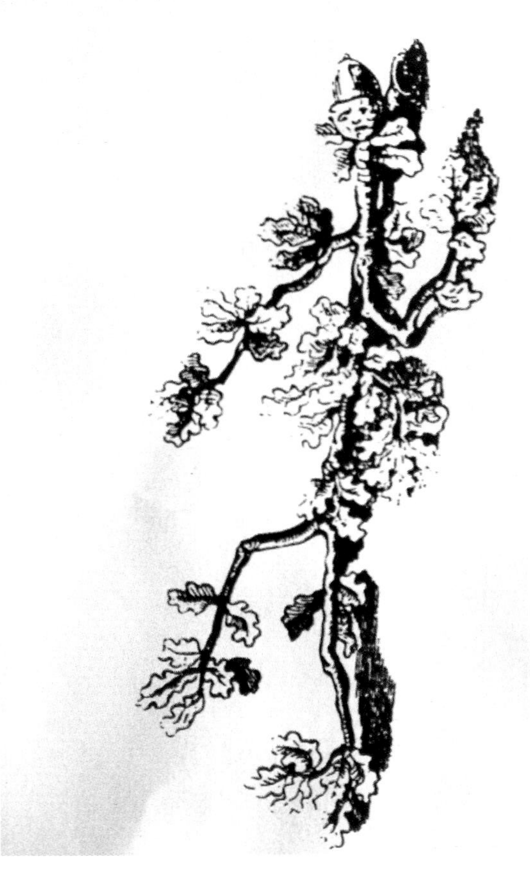

Fig. 3: Gribouille followed by Monsieur Bourdon (drawing by Maurice Sand in Sand, 1957)

Fig. 4: Gribouille changed in *rameau de Chêne* (drawing by Maurice Sand in Sand, 1957)

hitherto unknown world in which the self is confronted by, and divided between, opposing tendencies. This dream represents a battle between unconscious aggression, as symbolised by the bumblebee, and the power of reason that censures it.

The second sleep period amplifies the power of obscure forces that overwhelm the sleeper: the valet/spider represents both unpleasant characteristics that the child could have inherited from his parents (but which he refuses to acknowledge) and the energy he deploys in order to oppose his parents: a latent aggression that frightens and terrorises him. There is, therefore, a process of gradual realisation and awareness, which the third sleep period renders decisive. In this last sleep period, the internal combat becomes so violent that it is represented by a battle between the honeybee and the bumblebee. The shock of the third dream is sufficient for the child to understand that he can never gain his parents' love through aggression and that he must detach himself from them if he wishes to be happy.

The insect plays a fundamental role in this story, similar to that which it plays in *The Child Stranger.* The insect, with its demonic appearance, administers necessary lessons to Mankind, exactly as Master Ink did: the child 'takes' from existence (from his parents and from others), before the later stage of 'giving', which leads to the final sacrifice of self. This ambiguity in the insect's role, both useful and destructive, is also illustrated by the ambivalent role of the oak and fig trees, both of which are closely associated with the bumblebee and the honeybee in their initiatory function. They serve as the place where the spiritual and moral world meet and oppose each other through the power of the sleep which the tree induces.

Fig. 5: Alice and the caterpillar (drawing by Sir John Tenniel from Carroll, 1961)

■ *ALICE'S ADVENTURES IN WONDERLAND* AND *THROUGH THE LOOKING-GLASS*

In Lewis Carroll's *Alice's Adventures in Wonderland* (1865) and *Through the Looking-Glass* (1871), the insect and the forest are intimately linked in the world of our imagination. This link is demonstrated, for example, in the 'caterpillar's advice' given in *Alice's Adventures in Wonderland* which is comparable to her encounter with a gnat in *Through the Looking-Glass.* In her dream, Alice meets a large blue caterpillar, perched on a mushroom, with whom she is engages in a conversation about 'Who am I?' (Fig. 5). Both conversants are in intermediary stages: the caterpillar is neither larva nor butterfly and Alice herself is neither infant nor adult. The relationship between the two is reinforced by Carroll's insistence that they both measure seven centimetres: it is between seven and ten years that a child's mind develops 'a logical sense of classes and relations . . . applying to multiple sectors of experience such as numbers, space and time' (Greco, 1968: 214).

In *Through the Looking-Glass,* Alice is in a train which takes her to the fourth square in a game in which she has to cover eight squares. In this dream, an invisible gnat appears which, like the caterpillar, is capable of reading Alice's thoughts and questions her existence by proposing that she could lose her name. This is precisely what happens as soon as the gnat disappears, and she walks into the nearby forest. This time, the question of 'Who am I?' is not the consequence of a series of metamorphoses, but of the realisation of human impotence in a world in which names do not exist. Following this logic, when the fawn which she meets in the forest asks 'What do you call yourself?', Alice can only reply: 'Nothing, just now'. Alice finally discovers her identity *when she comes out of the forest*. Later, she travels through another forest from which she emerges as a queen (a person in her own right, in possession of reason). This second forest is located on the *seventh* square, demonstrating Carroll's insistence on seven years as the key age associated with the insect and with the forest. A place of psychic chrysalis, a closed place where the laws of the external world do not apply, a place where you are not yet yourself and yet where evolutionary processes continue; the forest is a labyrinth in which the Self reaches its definitive orientation.

■ THE INSECT, THE FOREST AND THE LABYRINTH

In his latest novel (*Shizukana seikatsu,*), Oé Kenzaburo (1990) focuses on a writer (himself) who has grown up in a valley surrounded by forests. Each time he goes through a moral or spiritual crisis, the writer experiences the need to retreat into places which he considers protective, i.e. those surrounded by trees. He is thus resorting to an archaic mental attitude by associating the forest with the 'Origin'; the origin of his life, no doubt, but also that of humankind.

Since the remotest times this link has been expressed by the mysterious image of the labyrinth[6], a representation of the 'Origin of Humanity' which presupposes a suspension of both time and space; a difficult journey towards wisdom, a journey that will cause the traveller to discover his identity once he has determined where he has been led astray. In this sense, the forest and the labyrinth are superimposed. Bachelard (1948: 217) has already stressed the extent to which the labyrinth represents a 'primary suffering, a suffering of childhood'. In addition, there are countless stories (such as Grimm's fairytales (Harrison, 1994: 246) in which a lost child must overcome trials and dangers in order to find his/her way out of the forest. It is in the forest that the insect (or animal) manifests itself and serves as a guide, which facilitates the subject's metamorphosis.

The symbolic role of the forest in Kenzaburo's work is found in another of his texts, *M/T to mori no fushigi no monogatari* (Oé, 1986). Here the narrator discovers 'Truth' inscribed in a drawing of a shoal of fish in the river in which he almost drowns; while time is suspended, the young boy discovers universal harmony (Siganos, 1993: 35). In effect, nothing exists except that which is plunged into an extremely complex network of relationships (i.e. the labyrinth); Man is not fully conscious of this network and only discovers 'Truth' at the moment when he can no longer express himself in language (Oé, 1986: 45). This discovery takes place in the forest, in a milieu favourable to the expression of the sacred, but also a feminine and even an amniotic space (Durand, 1969: 281)[7]. An encounter with an insect or an animal in the forest can therefore make one feel at the very 'centre', in harmony with creation. As Harrison (1994: 229) has stressed, 'the forests possess the psychic faculty of evoking memories of the past, they even constitute the metaphor of memory itself, in some way a metaphor which is bathed in the aura of lost origins'.

An isomorphism is thus created, as is clear in a story by Le Clézio (1978), *Peuple du Ciel,* in which a little girl receives a daily visit from the bees. She is in the habit of sitting at the top of a cliff overlooking her village in order to contemplate the sun, and the bees come and cover the entire face of this little blind girl, who then takes up with the insects a sort of wordless conversation. She is also stung, not by the bees but by a wasp, and the sting again plays an initiatory role for the moment of pain is followed by an ineffable state of communion with heaven (Le Clézio, 1978: 209).

In all these examples, the constant movement between attraction and repulsion is the instrumental process of fascination, which governs them. The archaic character of nostalgic fascination brings together the insect, the forest, and the labyrinth in a close symbolic relationship: all three offer to modern society a rereading of reality through metamorphosis, a changed state which would permit a return to the harmony of the 'Origin'.

Notes

1. Translated by Robert Griffiths.
2. Of the roughly five million species that exist in the world, nine-tenths are insects.
3. For other examples, see Berland (1942).
4. Here Jolles' (1930) classification is adopted which distinguishes between popular tales, the origin of which is collective, and 'learned' tales or those which are created by a particular author.
5. The first names chosen by Hoffmann are equivalent to each other in meaning besides being axiomatic of Christian morality: Felix (happiness) = Christlieb (love of Christ) and vice versa.
6. See Santarcangeli (1967), Lonegren (1991:93), Siganos (1993).
7. G. Durand stresses the extent to which the notion of centre is attached to the notion of 'navel', relating to a clearly masculine order of things, but also the extent to which this notion comprises 'obstetric and gynecological infrastructures' (Durand, 1960: 281). From this point of view the sacred mountains, like the lakes or the forests, are sacralised because of their spatially closed-off nature: 'It is for these uterine reasons that what renders a place sacred is above all its closed-off nature: islands with their amniotic symbolism, or forests where the horizon is closed in on itself. The forest is a centre of intimacy, as can also be the house, the grotto or the cathedral . . . the sacred place is indeed a cosmic form, much larger than the microscosm of an ordinary dwelling which is the archetype of feminine intimacy' (Durand, 1960: 281).

References

Abe, K. (1967: 1988). *Moyetsukita Chizu*, Shinchosha, Tokyo. *Le Plan déchiqueté*, Stock, Paris.

Aeppli, E. (1978): *Les Rêves et leur interprétation*, Payot, Paris.

Bachelard, G. (1948): *La Terre et les rêveries du repos*, rpt 1969, José Corti, Paris.

Berland, L. (1955): *Les Arachnides d'Afrique noire*, Institut Français d'Afrique Noire, Dakar.

Calame-Griaule, G. (1969–70): *Le thème de l'arbre dans les contes africains*, 16, 20, SELAF–Klincksieck, Paris.

Carroll, L. (1865): *Alice's Adventures in Wonderland, Alice au pays des merveilles*, rpt 1970, Aubier–Flammarion, Paris.

—— (1871): *Through the Looking-Glass; De l'autre côté du miroir*, rpt 1971, Aubier–Flammarion, Paris.

Chauvin, R. (1961): *Le Comportement social chez les animaux*, Presses Universitaires de France, Paris.

Chevalier J. et al., (1969): *Dictionnaire des symboles*, Seghers, Paris.

Durand, G. (rpt 1969): *Les Structures anthropologiques de l'imaginaire*, Bordas, Paris.

Frazer, J. G. (1966): *The Golden Bough: A Study in Magic and Religion*, (3rd edn.) Macmillan, St. Martin's Press, New York.

Greco, P. (1968): Enfance. Opérations et structures intellectuelles, in *Encyclopédia Universalis*, Paris, pp. 213–18.

Harrison, R. (1994): *Forêts. Essai sur l'imaginaire occidental*, Flammarion, Paris.

Hesiode (1928): *Les Travaux et les jours*, Belles Lettres, Paris.

Heusch, L. de (1972): *Le Roi ivre ou l'origine de l'Etat; Mythes et rites bantous*, Gallimard, Paris.

Hoffmann, E.T.A. (1832): L'Enfant Etranger, in Eugène Renduel, *Aux Enfants. Contes*, Paris, pp. 5–113.

Jolles, A. (1930): *Einfache Formen*, rpt 1972, Tübingen, *Formes simples*, Seuil, Paris.

Le Clézio, J. M. G. (1978): Peuple du ciel, in Gallimard, *Mondo et autres histoires*, Paris.

Lévi-Strauss, C. (1964): *Le Cru et le Cuit*, Plon, Paris.

Lhoste, J. (1979): *Des Insectes et des hommes*, Fayard, Paris.

Lonegren, S. (1991): *Labyrinths: Ancient Myths and Modern Uses*, Gothic Image, Glastonbury, *Les Labyrinthes. Mythes traditionnels et applications modernes*, Dangles, Paris.

Marchenay P. (1979): *L'Homme et l'Abeille*, Berger–Levrault, Paris.

Newman, L.H. (1965): *L'Homme et les Insectes*, Editions RST, Paris.

Oé, K. (1986): *M/T to mori no fushigi no monogatari, M/T et l'histoire des merveilles de la forêt*, rpt 1989, Gallimard, Paris.

—— (1990): *Shizukana seikatsu, une existence tranquille*, rpt 1995, Gallimard, Paris.

Pausanias (1965): *Description of Greece*, W. Heinemann Ltd, London, Harvard University Press, Cambridge, (W. H. S. Jones tr.).

Pindar (1922): *Pythiques*, trans. A. Puech, Belles Lettres, Paris.

Pliny the Elder (1952): *Histoire naturelle*, trans. E. Ernout, Belles Lettres, Paris.

Poignant, R. (1968): *Mythologie océanienne*, O.D.E.G.E., Paris.

Roberts, K. (1975): *The Furies*, Pan Books, London–Sydney.

Saint-John Perse (1972): *Amers, in Oeuvres Complètes*, Gallimard, Paris.

Sand, G. (1850): *Histoire du Véritable Gribouille*, rpt 1957, Le Club Français du Livre, Paris.

Santarcangeli, P. (1967): *Il Libro di labirinti.* Storia di un mito e di un simbolo, Vellechi edn., rpt 1974, Florence, *Le Livre des labyrinthes*, Gallimard, Paris.

Shinoda, C. (1994): *La Métamorphose des fées. Etudes comparatives des contes français et japonais*, Nagoya University.

Siganos, A. (1982): Sur Hoffmann et Sand: l'Histoire du véritable Gribouille et l'Enfant étranger; *Revue de Littérature Comparée,* 221: 92–95.

—— (1985): *Les Mythologies de l'Insecte*, Librairie des Méridiens–Klincksieck, Paris.

—— (1993): *Le Minotaure et son mythe*, Presses Universitaires de France, Paris.

Steiner, R. et al., (1958): *Triades: Abeilles, Hommes et Dieux*. Laballery, Clamecy.

Zahan, D. et. al. (1965): Aspects de la réincarnation et de la vie mystique chez les Bambaras, in *Réincarnation et vie mystique en Afrique noire*, Presses Universitaires de France, pp. 175–87.

Part IV

FORESTS AND FUTURE SOCIETY

CHAPTER 12

The Future Role of *Satoyama* Woodlands in Japanese Society

HIDEO TABATA

Introduction

Literally, *sato* means inhabited areas or villages, and *yama* means hills or mountains. *Satoyama* woodlands are areas of forest used by a particular village or community. Up until the 1960s, most Japanese woodlands (except those in very remote areas) had been used for many years to provide domestic and commercial firewood and charcoal, manure, edible wild plants, mushrooms, timbers, and other forest products. *Satoyama* woodlands were also repeatedly used for 'slash and burn' agriculture/shifting cultivation. These were generally of two types: those of deciduous oak composed mainly of *Quercus serrata*, and those composed of Japanese red pine (*Pinus densiflora*). Oak woodlands were regenerated by coppicing but red pine woodlands were usually reproduced by natural seeding or, sometimes, by afforestation. The Korean *dongsan* is almost identical to *satoyama* woodlands, but the coppice woodlands in England (Buckley, 1992) are not.

Coppicing to supply firewood and charcoal was the principal forestry practice in *satoyama* oak woodlands. However, most communities stopped using wood for fuel in the 1960s when propane gas became the standard domestic fuel even in rural areas. Charcoal production was remarkably reduced during the same period.

In rural Japan before the 1960s, wood for fuel and charcoal was obtained by cutting trees in *satoyama* woodlands every ten to twenty years. The shrub layer was very poor because shrubs were taken for firewood more often, although some shrubs were left as nutrient for the forests. The poor shrub layer and light forest floor provided good conditions for the growth of mushrooms and wild vegetables. Shrubs and leaf litter collected from the forest floor were used as manure for paddy fields but were replaced by chemical fertilisers in the 1960s. The poor soils in the Japanese red pine woodlands were good for the growth of some mushrooms such as *Tricholoma matsutake*.

Distributed in the *satoyama* area were bamboo forests, planted forests of *Cryptomeria japonica* and *Chamaecyparis obtusa*, and small patches of *Miscanthus sinensis* grassland for thatching. As mentioned above, the *satoyama* management was closely linked to the agricultural regime of paddy cultivation. The irrigation water for paddy fields or cultivated fields was supplied directly from *satoyama* woodlands or indirectly from irrigation ponds made in the *satoyama* area. It is difficult, therefore, to separate those woodlands from the agricultural lands located alongside them. Thus, *satoyama* is really an integrated landscape of woodlands, grasslands, bamboo forests, plantations, irrigation ponds, irrigation channels, paddy fields located next to the *satoyama* woodlands, the footpaths between them, and the like (Tabata, 1997a).

There are many differences between natural and *satoyama* woodlands. Within the latter there are various stands that are managed in different ways and are of different ages according to who owns them. *Satoyama* woodlands have a far simpler structure of stratification than their natural counterparts. Clear cutting is totally different from the gap formation in the natural forests. In *satoyama* woodlands, natural gap formation rarely happens and dead trees and fallen trunks are seldom found. Man-made gaps of some extent are formed at regular intervals according to the duration of clear cutting rotation. Species replacement after clear cutting is limited in *satoyama* oak woodlands because of coppice regeneration. However, the species composition shows stochasticity in *satoyama* red pine woodlands because of the regeneration by natural seeding.

The method and extent of utilisation of *satoyama* woodlands were altered periodically by the demands of the local economy, for example, wood fuel supply for the small scale iron industry, commercial firewood supply to the urban area, grazing in the woodlands, and the like. However, those woodlands came under severe pressure or were exploited when a cash economy was introduced to the rural areas in the seventeenth century, after which they increasingly deteriorated (Chiba, 1956, 1990; Ogura, 1992, 1996). This exploitation consequently caused drastic changes in the landscape from *satoyama* woodland to barren land in many areas.

The use of *satoyama* woodlands in modern times also appears to have been very exploitative, and barren areas were widely distributed throughout Japan even at the beginning of the Meiji era (in the mid-nineteenth century). However, the barren lands were somewhat restored when the pressure on the woodlands was decreased after the first fuel revolution in which coal and coal products were introduced as domestic fuel, though the recovery process of the vegetation was not recorded. A change in vegetation type has been taking place in the *satoyama* woodlands during these thirty years following the second fuel revolution in the 1960s.

Many woodlands still retain the basic features of well managed *satoyama* woodlands, and constitute still the most common type of vegetation in Japan, probably occupying over one-third of the total forest area. This coverage is despite the decline in *satoyama* woodland management and the related increase in abandoned or unmanaged *satoyama* woodlands over the past thirty years. It is also true that *satoyama* woodlands have been the most familiar natural environment for the Japanese people, who have enjoyed, for example, the changes in colour in autumn and springtime and on whose beauty much of Japanese culture is based.

The Current Situation

Satoyama woodlands have been destroyed by the expansion of cities, principally because over the past thirty years the woodlands have lost their commercial and forestry value. This however does not mean that the relationship between *satoyama* and humans has altered because of the recent changes in the Japanese lifestyle. The owners of *satoyama* sold them to developers without any consideration of their importance. Amongst other things, housing, factories, and golf courses have been built on *satoyama* woodlands. However, the benefit of these woodlands became clearly apparent when their destruction was followed by environmental pollution and disasters. Familiar plants and animals have been disappearing from our surroundings, and many of them, particularly inhabitants of the periphery of the *satoyama* woodlands, paddy field footpaths, artificial grasslands, wetlands, etc. are now vulnerable or endangered species (Tabata, 1997c).

Since biologists had not recognised the importance of the *satoyama* environment and woodlands, which had been disturbed by human activities and were, therefore, no longer in their natural state, *satoyama* has not been properly researched. We do not know the biological and ecological features of *satoyama*, and have only now started undertaking the necessary research on it in Japan.

The situation is very serious for those *satoyama* woodlands located on the hills of Pleistocene deposits as the topography of such land is easily modified into the flat lands by the use of construction machinery. The floral components of *satoyama* woodlands on the Pleistocene deposits are so rich that the destruction of those endangers more species than does that of the woodlands located in bedrock areas. Pleistocene deposits are principally distributed on the edge of plains and basins which also comprise heavily populated urban areas. Eventually woodlands that surround the urban areas tend to be destroyed for 'development' and the fragmentation of woodlands that development causes is another associated problem.

Paddy fields in small valleys surrounded by *satoyama* woodlands are usually less productive and were amongst the first to be abandoned. Footpath vegetation, which was grown and maintained in connection with the management of paddy fields, was the first to be destroyed whenever mowing took place. Delicate wetland and water plants (such as *Sanguisorba tenuifolia, Hosta longissima, Swertia diluta, Nymphoides peltata, N. indica,* etc.), which survived in or around wet paddy fields formed on the clay layer of Pleistocene deposits were lost because many irrigation ponds were abandoned. Another reason was that there was no fluctuation of pond water level due to abandonment of paddy cultivation in some *satoyama* areas (Tabata, 1997c).

After abandoning the management of *satoyama* woodlands which have prevented ecological succession, unmanaged *satoyama* woodlands are now changing into evergreen broad-leaved forests comprising evergreen oak species and *Castanopsis cuspidata*. In the near future we might well be deprived of the pleasures of the seasonal changes in colour that *satoyama* woodlands bring. This may also mean a change in the Japanese perception of nature as, apart from the southern part of Japan, the deciduous, broad-leaved forests have been the major element in the Japanese landscape. Indeed, Yasuda (1980) has argued that Japanese culture has been based on the deciduous broad-leaved forests or woodlands since Neolithic times.

In addition, the neglect of bamboo forests has led to the invasion of *satoyama* oak woodlands by bamboo. This replacement has had very serious repercussions for biodiversity as bamboo forests support few other plant species. Moreover, this will lead to soil erosion and the other ecological difficulties after flowering, as bamboo plants are monocarpic and die after flowering. Owners of *satoyama* woodlands have not complained about the invasion of bamboo forests because they are essentially interested in the ownership of the land rather than in the continuance and maintenance of the woodlands.

Biological Characteristics of *Satoyama*

■ FLORA

Satoyama woodlands in the central part of Japan are very similar to the lowland and hill forests of Korea and the north-eastern part of China, although the number of tree species is greater in China and Korea. *Quercus serrata* is dominant in Japanese *satoyama* deciduous oak forests while deciduous oak forests in Korea and north-eastern China comprise other oak species such as *Quercus aliena, Q. mongolica, Q. variabilis, Q. acutissima* and, in China, *Q. liaotungensis*. On the other hand, the principal species of evergreen broad-leaved forests are related to those in southern China and south-east Asia. These two types of forests appear to have a completely different history, although in Japan, *satoyama* woodlands are currently being replaced by climax forest of evergreen broad-leaved trees. In Korea and north-eastern China, where it is colder and drier than Japan, climax forests consist of deciduous broad-leaved trees containing several species of oak (Tabata, 1997b).

Our studies have shown that footpaths between paddy fields can be regarded as meadows, the importance of which has not been recognised yet. Although the width of the footpaths is narrow, they connect to each other to form large areas of artificially managed meadows that have been maintained since paddy cultivation began. Many species growing on paddy field footpaths are found in the meadows of north-eastern China, especially in those of Inner Mongolia. Amongst others, *Sanguisorba officinalis, S. tenuifolia, Gentiana squarrosa, Vicia pseudo-orobus, Lespedeza pilosa, Patrinia scabiosaefolia, Miscanthus sinensis, Phragmites communis,* and *Eupatorium lindeyanum* are all found in meadows of north-eastern China. *Lespedeza bicolor, Platycodon grandiflorum, Thymus serphyllum,* and *Agrimonia pilosa* growing on the fringes of *satoyama* woodlands are also found in the meadows of Inner Mongolia. *Inula britanica, Aster fastigiatus, Phragmites communis,* and *Sium suave* occur in Japanese wetlands or river beds and are also found in the meadows of north-eastern China. *Sophora flavescens* and *Echinops* which are found on artificial Japanese grassland also inhabits meadows in Inner Mongolia (Tabata, 1997b).

On the basis of the similarity of floristic components and physiognomy among Japanese, Korean, and north-eastern Chinese vegetation, it is possible to speculate that meadows similar to those of Inner Mongolia, and deciduous oak forests similar to those in Korea and north-eastern China, were found in Japan in the last-glacial period. *Satoyama* woodlands and paddy field footpath vegetation are therefore likely to be remnants of Inner Mongolian meadows and the deciduous oak forests of Korea and north-eastern China. This means that the artificially managed *satoyama* environment has maintained biodiversity of the last-glacial period even in the Holocene together with warm elements.

■ MUSHROOM FLORA

In Kyoto (central Japan) the mushrooms of the *satoyama* woodlands (*Pinus densiflora* and *Quercus serrata* woodlands) tend to emerge in autumn. They are basically boreal and are distributed in the temperate zone from Japan to the Eurasian continent through Korea. They are represented by *Lyophyllum shimeji, Boletopsis leucomelas, Rozites caperata,* and *Tricholoma matsutake.* On the other hand, the evergreen broad-leaved forests are dominated by tropical or subtropical mushrooms like *Hymenogaster arenarius* and *Amanita perpasta* whose fruit bodies emerge in midsummer.

Tropical or subtropical mushrooms grow in bamboo forests. *Dictyophora indusiata, Cryptoderma asparatus,* and *Psilocybe lonchophorus* are represented here and fruit bodies are found in hot summer. These species are generally confined to bamboo forests even though red pine and oak woodlands are located very closeby. Some species of mushroom are common to both bamboo and evergreen broad-leaved forests (Yokoyama and Sakuma, 1997).

■ ANT FAUNA

In Kyoto, *satoyama* woodlands are very rich in ant species because of the mosaic structure of the environment. It is remarkable that rare ant species, especially submerged species, are found in *satoyama* woodlands. They are represented by *Discothyrea sauteri, Monomorium triiale, Pentastruma canina, Epitritus hermerus* and *E. hirashimai*. Boreal species like *Dilchoderus sibiricus* inhabit oak woodlands whereas tropical or subtropical species such as *Epitrutus hexamerus* and *E. hirashimai* are found in bamboo forests. Evergreen broad-leaved forests are also inhabited by tropical or subtropical ant species (Onoyama, 1997). Just as the association between fungi and the host plants has been roughly maintained, so has the biological association between particular ant species and vegetation. This association has been maintained at least since the last ice age.

■ ECOLOGY OF POND-INHABITING INSECTS

A comparison of the behaviour of various pond-living insect species was conducted in a pond in Nara, central Japan (Hibi, 1993; Hibi and Yamamoto, 1997). In early summer, most individuals of these species disappeared from the pond where they hibernated, though some individuals of *Notonecta triguttata* stayed throughout the year (Fig. 1). There were some differences in the time the species of insect left the pond and their duration of absence. *Cybister brevis* was the earliest to leave the pond and the last to return. By marking the insects it was shown that in summer they travelled to paddy fields 1–1.5 km away to reproduce. These insects cannot survive unless there are submerged paddy fields within a distance of 1–1.5 km from the pond; a combination of paddy fields and ponds is therefore essential for their survival.

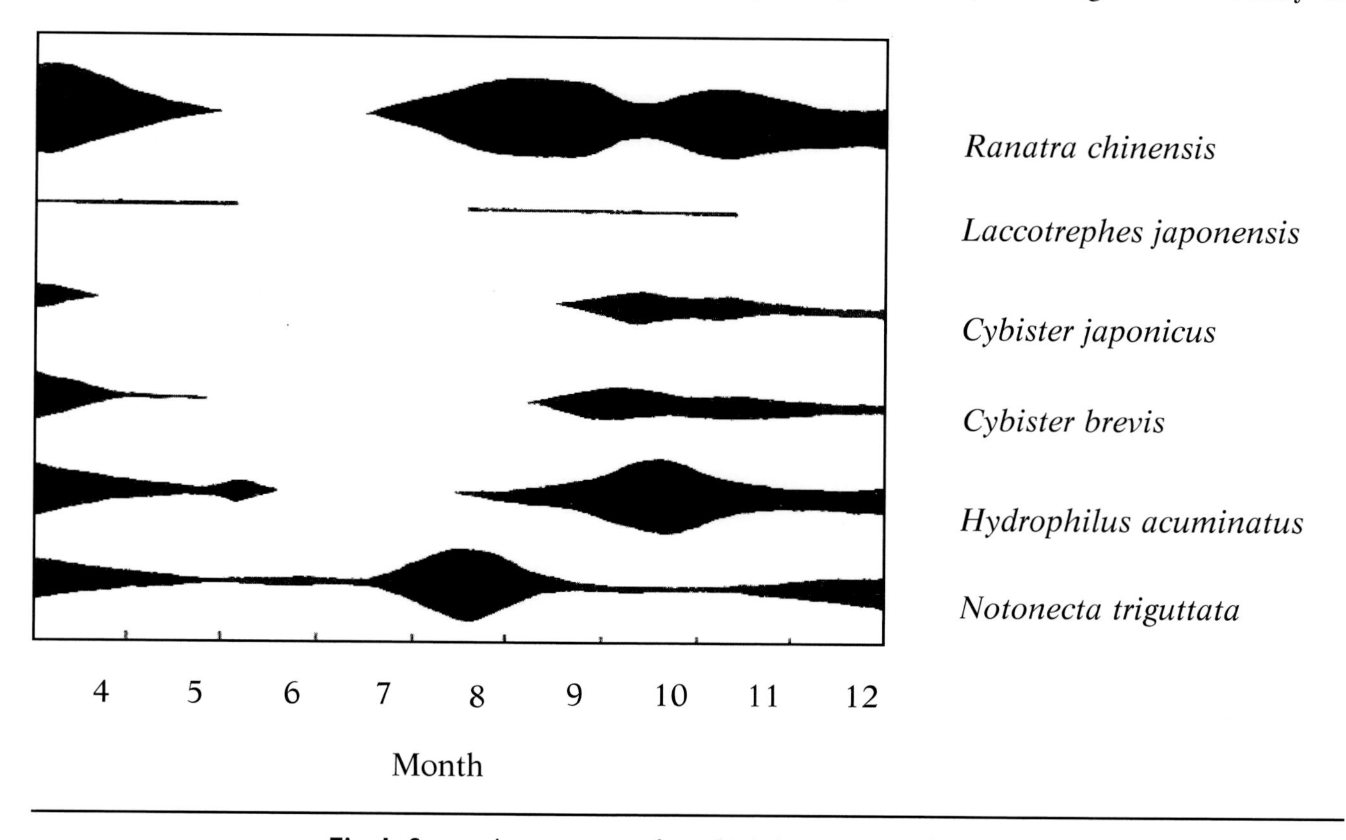

Fig. 1: Seasonal occurrence of pond-inhabiting insects (Hibi, 1993)

■ ECOLOGICAL INTERCONNECTION

A good example of the ecological interconnection in the biological community is provided by *Sophora flavescens*, which grew abundantly in grasslands and was used for thatching and pasture for domestic animals and *Shijimiaeoides divinus* which only feeds on *Sophora*. As grasslands decreased in area and number, the population of *Sophora* decreased and this butterfly has now become an endangered species. It is perhaps ironic that both these species now survive on the practice fields of the Japanese Self-Defense Forces (Ishii, 1997).

Because trees and shrubs were removed from woodlands, the practice of coppicing reduced the fauna and flora in British coppice woodlands (Evans and Barkham, 1992). Although this observation may have important ecological implications for *satoyama* woodlands, similar studies have not been undertaken in Japan.

The Relationship Between *Satoyama* and People

It is important to note that *satoyama* woodlands should be maintained both for the conservation of biodiversity and an amenity for people, even though *satoyama* woodlands have now lost their economic value and there is little hope of changing the attitude of those who own them. *Satoyama* woodland management involved artificial disturbance of some scale in intervals of clear cutting rotation, which provided various habitats to the inhabitants of the areas. Charcoal production is in this sense the most attractive practice for the conservation of *satoyama* woodlands because the clear cutting for charcoal production can produce a mosaic structure in the woodlands. The problem is the cost of charcoal production. Charcoal can be put to various uses in the environment industry, which provides a sustainable use for it as an alternative to its use as fuel. There is a hope of developing large-scale, cheap charcoal production (Tabata, 1997d).

Small-scale electrical power production, which uses wood for fuel, is another new method that both uses woodland resources sustainably and produces a mosaic forest plan. It is also challenging as it depends on sources of renewable energy with a minimal detrimental impact on the environment. Besides, the heat efficiency will be about 85% in a small-scale biomass energy system when the heat is used for district heating and cooling as opposed to about 35% in large-scale electric power production in which heat is wasted, including the loss during electric transmission (Koike, 1997; Tabata, 1997d).

Who is responsible for the management of *satoyama* woodlands? I believe that so long as the owners of these woodlands continue to ignore their management we will need to introduce the concept of 'social commons', establishing that woodlands are for the use of the community as a whole (in the same way as schools, hospitals, public libraries, bridges, roads etc.). One way of achieving this may be to assign the care of private woodlands to local public forestry corporations or local forest owners associations in the form of contractual agreement with the owners in a transitional stage to full public management. Eventually, local governments should be responsible for *satoyama* woodlands although some management should be undertaken by the central government as these areas have many regulatory functions relating to the environment. In addition, I believe that only professionally trained forest workers belonging to governmental forest agencies, local public forestry corporations, and/or local forest owners' associations should be engaged in *satoyama* forestry.

Recent *satoyama* research has shown that this environment, as well as the adjacent agricultural environment, is necessary for the survival of many plants and animals, suggesting the kind of landscape arrangement that should be conserved in the future for the maintenance of biodiversity. Such a landscape would also constitute an amenity for the Japanese people.

Many perennial herbaceous species found on paddy field footpaths are now vulnerable even though the footpaths are well managed and herbicides are not used. For example, it is no longer easy to find *Gentiana scabra, Platycodon grandiflorum,* or *Eupatorium fortunei*. It is also difficult to find them on the footpaths of abandoned paddy fields. Why are familiar autumn flowering herbaceous plants vulnerable on well-managed footpaths? It is possible that their disappearance is related to the introduction of early-ripening varieties of paddy rice. Farmers used to mow the footpaths before they harvested the paddy fields. The last mowing at the end of September, therefore, caused the most serious damage to autumn flowering herbaceous plants. Before the introduction of early-ripening varieties, these plants could flower, seed, and reserve the assimilation products before the last mowing. Moreover, while almost all rice varieties today are early-ripening species, farmers used to cultivate late-ripening, mid-season-ripening, and early-ripening varieties. The abandonment of paddy fields located next to *satoyama* woodlands seriously affected herbaceous plants because the footpaths also became neglected. It is thus essential to restore paddy fields as an agricultural environment for environmental sustainability, landscape and biodiversity.

There are some delicate political and economic problems here, and we may need a political decision to restore the abandoned fields. We could introduce concepts such as the 'Kleingarten' in Germany (Toshitani and Wada, 1994) and open up the fields to the public for the restoration of the agricultural environment under a contract between the owner and the new cultivator. Many city dwellers and active retired people are seeking land to undertake agricultural activities.

In conclusion, it is also important to recall the intimate relationship between human beings and nature, especially woodlands/forests. Human sensibility is developed in the cradle of nature, especially woodlands/forests. In the forests there are various kinds of sounds, smells, colours, and shapes which cannot be artificially created. We can experience and develop all sorts of senses in the forest. This experience is very important in the infant stage of ontogenetic development of human beings. We may need to increase the opportunites for children to develop a contact with nature and spend time in the natural settings of woodlands/forests for healthy development. Woodlands/forests are an essential human amenity and important for our well being and healthy development because they generate a sense of peace and amity when we are there and even when we view a landscape with forests. It is therefore imperative for the present and future generations that we preserve *satoyama* in our living environment.

References

Buckley, G. P. (ed.), (1992): *Ecology and Management of Coppice Woodlands*. Chapman & Hall, London.

Chiba, T. (1956): *Studies on Barren Hills*. Gakuseisha, Tokyo (in Japanese).

—— (1990): *Studies on Barren Hills* (rev. ed.), Sosiete, Tokyo (in Japanese).

Editing Committee of Natural Geography of China, Chinese Academy of Sciences. (1988): *Natural Geography of China*, Vol. 2, Science Publisher, Beijing. (in Chinese).

—— (1980): *Vegetation of China*, Science Publisher, Beijing (in Chinese).

Evans, M. N. and J. P. Barkham, (1992): Coppicing and Natural Disturbance in Temperate Woodlands: A Review. pp. 79–98, in G. P. Buckley (ed.), *Ecology and Management of Coppice Woodlands*, Chapman & Hall, London.

Hibi, N. (1993): Nature of Satoyama, *Gonta*, 3 (2): 4–5, 3 (3): 2–3 (in Japanese).

Hibi, N. and T. Yamamoto (1997): Pond-Inhabiting Insects in *Satoyama*, pp. 78–81, *in* H. Tabata (ed.), *Satoyama and Its Conservation*, Hoikusha, Osaka (in Japanese).

Ishii, M. (1997): The Change of Butterfly Fauna in *Satoyama*, pp. 126–132, in H. Tabata (ed.), *Satoyama and Its Conservation*, Hoikusha, Osaka (in Japanese).

Koike, K. (1997): *Satoyama* Woodlands Can Supply Clean Energy of Carbon Dioxide Emission Zero. pp. 194–6, in H. Tabata (ed.), *Satoyama and Its Conservation*, Hoikusha, Osaka (in Japanese).

Ogura, J. (1992): *History of Human Beings and Landscape*, Yuzankaku, Kyoto (in Japanese).

—— (1996): *Life of the Japanese Viewed from the Vegetation in Meiji Era*, Yuzankaku, Kyoto (in Japanese).

Onoyama, K. (1997): Characteristics of Ant Fauna of *Satoyama*, pp. 66–9, in H. Tabata (ed.), *Satoyama and Its Conservation*, Hoikusha, Osaka (in Japanese).

Red Data Book Committee, (1989): *Red Data Book of Japan*, Japan Association of Natural Conservation, Tokyo.

Tabata, H. (1997a): What Kind of Nature is *Satoyama*? pp. 6–9, in H. Tabata (ed.), *Satoyama and Its Conservation*, Hoikusha, Osaka (in Japanese).

—— (1997b): Flora of *Satoyama*. pp. 35–43, in H. Tabata (ed.), *Satoyama and Its Conservation*, Hoikusha, Osaka (in Japanese).

—— (1997c). Loss of Irrigation Ponds, Paddy Fields and Grasslands, and Endangered Species, pp. 118–125, in H. Tabata (ed.), *Satoyama and Its Conservation*, Hoikusha, Osaka (in Japanese).

—— (1997d): Our Proposal From the Viewpoint of *Satoyama* Research, pp. 164–75, in H. Tabata (ed.), *Satoyama and Its Conservation*, Hoikusha, Osaka (in Japanese).

Toshitani, N. and T. Wada, (ed.), (1994): *A Vision of Realization of Japanese Kleingarten*, Gyosei, Tokyo (in Japanese).

Yasuda, Y. (1980): *Introduction to Environmental Archeology*, NHK Publishing, Tokyo (in Japanese).

Yokoyama, K. and D. Sakuma, (1997): Characteristics of Mushroom Flora of *Satoyama*, pp. 50-57, in H. Tabata (ed.), *Satoyama and Its Conservation*, Hoikusha, Osaka (in Japanese).

CHAPTER 13

The Past, Present and Future of Medicinal Plants in Central America with Special Emphasis on Those For the Treatment of Parasitic Diseases in Guatemala

JUN MAKI

During the International Botanical Congress held in Yokohama, Pignatti (1993) issued a global botanical warning. This warning pointed to the great decline in tropical forests as, amongst other factors, due to population growth, a worldwide economic crisis, and environmental pollution. Takido (1993) stated that medicinal plants, which have diminished with the global shrinkage of forests, should be protected.

This situation also holds true in Central American countries such as Guatemala. Here, the percentage of forest cover has been greatly reduced. Today, less than 50% of Guatemala is covered by forest compared to about 90% fifty years ago (Caceres, 1995). Symbolic of the destruction of Guatemala's natural habitat is the decline in population of its national bird, the quetzal. The Guatemalan national tree, *Ceiba pentandra* is also now rare, having been felled to make way for pasture. It is regrettable that medicinal plants have also been diminishing with the decrease in forest area (Caceres, 1995). Medical and pharmaceutical scientists must take action before medicinal plants become extinct.

Parasitic Diseases and Medicinal Plants in Guatemala

Central America is the geographical and cultural region where the Maya Empire flourished for several centuries. The confluence of the Atlantic and Pacific Oceans, the rich lowland rain forest, and the highlands, which represent the last portion of the Sierra Madre, are determinant factors for the high biodiversity of this region. With the onset of the colonial era, several expeditions were sent to North and South America to locate, describe, and collect plants of medicinal importance from this region. Nevertheless little was studied about Mesoamerica. According to Caceres (1995), the density of medicinal plants in Costa Rica, Central America, is amongst the highest in the world, suggesting the possibility that the inhabitants have been familiar with them. In this paper, attention will be particularly focused on Guatemala, one of the Central American countries for which the experts of the Japan International Cooperation Agency (JICA) have been working in tropical disease studies under the chairmanship of Prof. I. Tada (Faculty of Medicine, Kyushu University). Since 1993 I have spent two to three months each year in that country as a JICA expert in medical sciences under the guidance of team leaders, Drs K. Ogata and Y. Tabaru and the encouragement of Prof Y. Ito (Kitasato University School of Medicine).

Guatemala is an interesting country with a Mayan and Spanish historical background, as I have shown in my presentations (Maki, 1995), attracting many tourists. Apart from its culture and history, it holds attraction too for natural scientists from zoological and botanical viewpoints. From November to May, it has a dry season and for the other months it is rainy. There are over 30 volcanoes and the highest mountains rise 4,000 m above sea level in the total territory of 110,000 km^2 In other words, Guatemala, located in a tropical region, is a mountainous country and several ecological and cultural circumstances coexist to make it of tremendous interest.

One of the biological characteristics of Guatemala is the presence of a wide variety of parasites and their vectors responsible for human infectious diseases, especially intractable ones such as malaria, Chagas disease, leishmaniasis, and filariasis (*Onchocerca volvulus*) are rampant. Most of them even now remain incurable diseases notwithstanding the efforts in the west to develop and synthesise effective drugs to treat them. The Guatemalan people have been suffering from obstinate parasitoses. When this problem is overcome in the future, Guatemala is certain to be a much more attractive country for international tourists as it is a country that can boast a picturesque and magnificent natural environment.

The other characteristic is the abundance and diversity of medicinal plants that have been traditionally used for treatment of the above mentioned ailments from Maya and Spanish times. Drugs from indigenous trees and grasses have played a significant role in safe and effective treatment of diseases, including those by parasites in Guatemala. Such drugs have, besides the question of safety, the advantages of easy availability and low cost. For instance, the villages around the volcanic lake Atitlan, so remote from the capital, are well known for their traditional daily use of medicinal plants (Caceres, 1995). JICA has attached much importance to the problem in Guatemala where infectious diseases are still rampant, and various kinds of medicinal plants have traditionally been utilised.

■ SCIENTIFIC ACTIVITIES IN THREAT OF A DECREASE IN MEDICINAL PLANTS IN THE FORESTS OF GUATEMALA

There are several kinds of intractable parasitic diseases in Guatemala, which are by no means easy to cure. Even if modern drugs had been synthesised elsewhere, such drugs would have been more or less inaccessible to the people in the rural regions and their cost would have been exorbitant by local standards. If plant-origin drugs with high efficacy and without any severe side effects are readily available to inhabitants living so remote from the towns, this would undoubtedly be an enormous boon to them, and the signs are that the available medicinal plants can form the basis of inexpensive and safe treatment of parasitic diseases which have so far been resistant to conventional treatment. There are however some problems in the utilisation of plant-origin drugs.

While parasitic diseases are still rampant causing immense suffering, a great proportion of the natural resources, including medicinal plants, have been disappearing in recent years in Guatemala. One of the urgent necessities in the present situation is the domestication of useful wild medicinal plants so that they do not disappear forever. This will be described later in somewhat greater detail.

Another problem is that folk medical doctors are aging, are not being replaced and therefore information on traditional herbs has been hard to obtain in recent times. There is an urgent need to compile a description of the plants used. Molina (1992) reported medicinal plants traditionally used among Guatemalan people, some of them considered to be useful for the treatment of parasitoses. This description does not however provide scientific names. The plants traditionally utilised to cure parasitoses are described in greater detail in the next section.

At the same time, we have to consider that almost all the plant-origin drugs in Guatemala have hitherto been administered to patients based on empiricism or popular beliefs. With the exceptions of quinine from *Cinchona* spp. as an antimalarial drug and ascaridol from *Chenopodium ambrosioides* as anti-round-worm drug, the use of crude drugs against parasites is without a scientific background. The efficacy of the drugs should be scientifically validated for reasonable administration to patients. It goes without saying that non-effective drugs should not be administered. The experimental studies on this point will be described in this paper.

■ INFORMATION ON MEDICINAL PLANTS TRADITIONALLY USED FOR THE TREATMENT OF PARASITIC DISEASES IN CENTRAL AMERICA.

The scientific validation of plant-origin drugs in antiparasitic effectiveness is impossible unless ethnobotanical information is available. I and my co-worker Prof. A. Caceres (University of San Carlos, Guatemala) surveyed the available ethnobotanical literatures which describe medicinal plants believed to be useful for the treatment of parasitic infections (Maki and Caceres, 1993). The ethnobotanical information was collected on the possible medicinal plants popularly used in the Mesoamerican region against parasites. About 280 species of plants were found to be used, or believed to be effective, against parasites in Mesoamerica. A somewhat detailed review was prepared and presented (Caceres et al.,1993; Maki and Caceres, 1993; Caceres,1995). Their family names are as follows:

Acanthaceae, Agavaceae, Amaranthaceae, Anacardiaceae, Annonaceae, Apocynaceae, Araceae, Aristolochiaceae, Asclepiadaceae, Bignoniaceae, Bixaceae, Boraginaceae, Bromeliaceae, Burseraceae, Cactaceae, Cappaveraceae, Caricaceae, Chonopodiaceae, Commelinaceae, Compositae, Convolvulaceae, Crassulaceae, Cruciferae, Cucurbitaceae, Elatocarpaceae, Equisetaceae, Euphorbiaceae, Gencianaceae, Graminea, Guttiferae, Hippocrateaaceae, Hydrophyllaceae, Labiatae, Lauraceae, Leguminosae, Liliaceae, Loganiaceae, Malpighiaceae, Malvaceae, Meliaceae, Menispermaceae, Monimiaceae, Moraceae, Moringaceae, Musaceae, Myristicaceae, Myrtaceae, Nyctaginaceae, Oxalidaceae, Palmae, Papaveraceae, Passifloraceae, Phytolaccaceae, Piperaceae, Plantaginaceae, Plantaginaceae, Polygonaceae, Portulaccaceae, Punicaceae, Rhamnaceae, Rhizophoraceae, Rosaceae, Rubiaceae, Rutaceae, Salicaceae, Sapindaceae, Sapotaceae, Selaginellaceae, Simaroubaceae, Smilacaceae, Solanaceae, Sterculiaceae, Theophrastaceae, Umbelliferae, Urticaceae, Valerianaceae, Verbenaceae, and Zingiberaceae.

It was difficult to describe the exact parasite species because most of the expelled worms are recognised usually by inhabitants not by parasitologists. However, they are classified into three categories. The first group related to plants used for the purpose of expelling helminth parasites, the second to plants used for the treatment of infections with protozoal parasites, and the third to those used for the other parasitic infections. About 70% of the plants listed are said to be useful for the treatment of helminthic infections, about 61% for the treatment of protozoal diseases such as dysentery (amebiasis), fever with chills (malaria), vaginitis (trichomoniasis), and chronic ulcers (1eishmaniasis). About 10% of the plant species are useful for the treatment of other parasitic infections.

■ EXPERIMENTAL STUDIES ON EFFECTS OF PLANT-ORIGIN DRUGS AGAINST PARASITES

As mentioned above, there still remain a number of species of causative agents responsible for obstinate infectious diseases. The present authors (Maki and Caceres, 1993) examined the efficacy of medicinal plant extract against these parasites, for instance the dengue virus, Iarval *Taenia solium* and *T. cruzi*.

The plants believed to be useful for the treatment of parasitic diseases are worthy of being examined for their scientific validity by comparison of experimental (medicated) and control (non-medicated) groups. In this way, studies have been carried out to clarify the spectrum of the effective crude drugs since the collection of the information. Special emphasis has been given to studies on the efficacy of traditional plant extracts against *T. cruzi*, a causative agent of Chagas disease (Caceres et al., 1996). Some of the drugs were found to be effective, others demonstrated not to be so, as will be outlined later.

Chagas disease is endemic not only to South American countries but also to Central America including Guatemala. It is an ailment of grave concern in rural areas like Santa Maria lxhuatan, Guatemala. Unfortunately, it has been a somewhat orphan disease notwithstanding the fact that the Guatemalan people have widely suffered from this protozoal disease. Few studies or actions taken for the control of this infectious disease have been reported; no chemotherapy using synthesised drugs has been established so far. To overcome Chagas disease, therefore, there will hopefully be a way of treating it with medicinal plant extracts.

Drugs from indigenous plants are to this day expected to play an important role in the treatment of various kinds of infectious diseases in Guatemala. However, no medicinal plants have been reported to be useful for the treatment of Chagas disease, so we could not help testing the extracts from the plants believed to be effective against protozoal infection (Maki and Caceres, 1993; Caceres et al., 1995). This contribution describes the trial, supported by JICA, to find useful plants in the future.

■ COLLECTION OF MEDICINAL PLANTS TO TEST THE EFFICACY OF *T. CRUZI*

Medicinal plants traditionally used for the treatment of protozoal diseases were collected on the basis of ethnobotanical information and the literature mentioned earlier in this paper. The plants were mainly harvested in the Mazatenango area located about 160 km away to the west of Guatemala City. A short trip from Guatemala city via Palin to the Mazatenango area was planned and made by Caceres and me in mid-February 1994 (Maki and Caceres, 1994). In the afternoon we left the capital for Mazatenango by car. We stopped by Palin where merchants were selling many kinds of fruits under a huge tree of mythical importance (*Ceiba pentandra*). The twigs and foliage cover a very wide area, giving the fruits on the ground a form of protective shade.

As we drove on, we lost sight of pine trees. We were now on the lowlands. In the course of time automatically watered sugar-cane fields came into sight, but we noticed that the cane had largely been cut because it was the season for harvesting sugar-cane. The cane was piled up on trucks to be transported, some being scattered on the roads. The production of sugar-cane per hectare in the Guatemalan lowlands is higher than that in any other country. This region is the golden belt of Guatemalan wealth, but the wages paid to workers are said to be quite low (up to 60 quetzals a day). The poverty of this region is in reality, in my view due to the fact that it is no longer one of tropical rain forest with an abundance of medicinal plants that it probably once was. The sugar-cane cultivation was started and sustained at the cost of these plants. A farming landscape hove into view. On the farms, a few *Ceiba* trees remain, a reminder of the former forests. The agro–industry producing sugar, cotton, and rubber is important for Guatemala, but at the same time the plant resources should be protected before they become extinct. One of the most important resources in Guatemala should be the forests and grasses including medicinal plants.

We crossed three bridges which had earlier been destroyed by guerrillas, two now reconstructed, the other tentatively so. When we were going to cross it, we were involved in a traffic jam for nearly an hour. In fear of a possible attack by somebody, we accelerated in order to reach our destination before sunset. At about six o'clock, we arrived at a house in a finca safe and sound. This region, about 400 m above sea level, is endemic with malaria and dengue which are transmitted by mosquitoes. The following day we went to collect medicinal plants for study.

The plants (parts of the plants examined), *Bixa orella* (leaves), *Byrsonima crassifolia* (bark), *Neurolaena lobata* (leaves), *Solanum nigrescens* (leaves), *Tagetes lucida* (leaves) and other species were harvested in Mazatenango.

Details of the plants follow (Maki and Caceres, 1993, Maki et al., 1995) and their believed efficacy, apart from that against parasitic diseases is briefly described.

- *B. orellana* = a tree 3–9 m tall locally called Achiote. It is found from Mexico to Bolivia at 0–100 m above sea level. In Guatemala it is grown to obtain dye materials. The maceration and decoction of the seeds are apparently used for weakness, diabetes, flu, venereal diseases, diarrhoea, jaundice, hemorrhoids, and olygouria.
- *B. crassifolia* = a tree 1–10 m tall distributed from Mexico to South America 0–1,800 m above sea level. The bark decoction is consumed for toothache and gonorrhoea. The bark infusion is consumed for diarrhoea, dysentery, and stomachache.
- *N. lobata* = The local name is Tres puntas. It is a tree about 3 m high with small yellow flowers distributed from Yucatan to northern Venezuela raging from sea level to 1,500 m elevation. The infusions of the leaves are popularly used to relieve pain, nervous weakness, anaemia, diabetes, low blood pressure, and healing chronic ulcers.
- *S. nigrescens* = a herb 0.5–2 m tall. Native from Mexico to Coastarica, it is found in bushes from 1,500–3,900 m above sea level. It is said to be useful for the treatment of dermatomucosal diseases, wounds, sores, pustules, ulcers and vaginitis, asthma, tonsilitis, cirrhosis, diarrhoea, toothache, scurvy, constipation, gastritis, swellings, meningitis, neurosis, high blood pressure, urinary retention, rheumatism, whooping cough, and gastric ulcer. The leaves are eaten by people recovering from various diseases.
- *T. lucida* = The local name is Pricon. It is an aromatic perennial herb 30–90 cm high with yellow flowers distributed from Mexico to northern Central America at 1,000 to 2,000 m above the sea level. In Guatemala, the decoction, infusion, or tincture of the leaves and flowers are used for spasmodic pains, diarrhoea, gastritis, and rheumatism.

■ BIOASSAY OF EFFECTS OF THE PLANT EXTRACTS ON *T. CRUZI*.

The plants collected were washed with water and shade dried. Test materials were extracted with dichloromethane, ethanol, and water, and concentrated by a rotavapor or lyophilisation and stored in a vacuum desiccator. A mixture of 0.1 ml epimastigote (Tulahuen strain) suspension in LIT culture medium and 0.1 ml extract-LIT medium was agitated in small holes of a plate and incubated at 28° C for 48 hours. The negative control was free from any extract. Preliminary results (Maki et al., 1995; Caceres et al., 1995) based on the microscopic observation indicated that the extracts from about one-third of the plant preparation so far examined were thought to be active against *T. cruzi*. For example, the extracts (final concentration of the extract: 25 mg/ml) from *N. lobata* and *T. lucida* were shown to be highly effective in reducing the number of parasites. The extract from *N. lobata* was highly effective even at a lower concentration (0.5 mg/ml) while *T. lucida* was shown to be lower in efficacy. *In vivo* studies are now in progress.

Incidental to the study of the effects of the extracts on *T. cruzi*, their effects on a number of microbes of public concern and medical importance were also studied (Caceres et al.,

1995). Their activity against bacteria (*Pseudomonas aeruninosa, Salmonella typhi,* and *Staphylococcus aureus*), yeast (*Candida albicans* and *Cryptococcus neuformans*), and fungi (*Aspergillus flavus* and *Microsporum gypseum*) were screened *in vitro* by dilution procedures. According to the result the extracts from *B. crassifolia, N. lobata* and *T. lucida* were notable for their efficacy. Their validity, based on scientific evidence, will suggest that the plants are beneficial to the well-being of the Guatemalan people.

Conclusion

My colleague in the JICA study, A. Caceres, stayed in Japan from October to December 1995. During his stay he visited a number of medicinal plant gardens for the protection and cultivation of medicinal plants in Guatemala, remembering that Guatemalan plants are now endangered. In the near future, specialised medicinal plant gardens will hopefully be set up in Guatemala supported by international funds to enable the plants generated there to be cultivated in the fields again.

The information on medicinal plants believed to be useful for the treatment of parasitic diseases in Guatemala has been collected to the extent cited above and will hopefully be presented in an international journal. Based on the information, experimental studies will be undertaken by a group of researchers, parasitologists, and pharmaceutical scientists.

Caceres and Maki have been studying the effects of medicinal plant extracts against *T. cruzi* mentioned here. We are now expanding our examination to other parasites too. In this way a spectrum of plant extracts will have been examined.

In conclusion it must be said that the protection of medicinal plants that are in danger of becoming extinct should go with a scientific recognition of the utility of the plants as a traditional cultural heritage of Guatemala. The good example of this has been the protection and cultivation of quinine trees following the collection of information available about them from the past.

I would like to express my sincere gratitude to Guatemalan and Japanese co-workers and advisers in the cooperative studies financially supported by JICA.

References

Caceres, A. (1995): *Search of Antimicrobials From Natural Products in Guatemala*. Seminar at the Kitasato Medical Society, 6 November.

Caceres et al., (1995): *Anti-Trypanosoma cruzi Activity of Extracts from Medicinal Plants Harvested in Guatemala*, The 37th Annual Meeting of Japanese Society of Tropical Medicine, Abstract p. 60.

—— (1995): *Antimicrobial Activity of 13 Native Plants Used in Guatemala for the Treatment of Protozoal Infections*, The 37th Annual Meeting of Japanese Society of Tropical Medicine, Abstract p. 61.

Caceres, A., B. Lopez, J. Maki, and I. Tada (1996): Anti-*Trypanosoma cruzi* Activity of Extracts from Native Medicinal Plants in Guatemala for Protozoal Infections, *Japanese Journal of Tropical Medicine and Hygiene,* 24: 59–60.

Maki, J. (1995): Maya Civilization and Forests: Endemic Parasitic Diseases and Traditional Medicinal Plants in Central America with Special Emphasis on Those in Guatemala, Forests, Myths and Civilizations, WA NO KUNI, An International Symposium, 14–17 December, 1995, Nara, Japan.

Maki, J. and A. Caceres (1993): *Preliminary Studies on the Effects of Plant-origin Drugs against Parasites, Especially Trypanosoma cruzi in Guatemala*, A report on the project of research for the control of tropical diseases, Japan International Cooperation Agency, p. 21.

—— (1994): *Experimental Studies on the Efficacy of Extracts From Medicinal Plants in Guatemala Against Trypanosoma cruzi*, JICA Report on Investigation of Tropical Diseases in Guatemala, p. 14.

Maki, J., A. Caceres,, M. B. Lopez, and S. Gonzalez (1995): *Fundamental Studies for the Development of Chemotherapy of Chagas Disease with Extracts from Medicinal Plants in Guatemala*, Progress Report submitted to JICA, p.15.

Molina, P. A. (1992): La Medicina Tradiconal o popular de Guatemala, Enfernredades Tropicales en Guatemala, Guatemala, JICA, pp. 54–70.

Pignatti, S. (1993): For A Global Botanical Warning, *XV International Botanical Congress Daily Bulletin,* 29 August 1993, Yokohama.

Takido, M. (1993): A Warning for the Decrease of Medicinal Plants on the Earth, *Farmacia,* 29: 743.

Part V

CONCLUSION

Forest and Civilisations

YOSHINORI YASUDA

Myths of the Forest

The nineteenth century was a period when ancient civilisations were discovered. H. Schliemann (1822–90), attracted by the ancient Greek epics, the *Iliad* and the *Odyssey*, made the greatest discovery in the history of mankind. This was clearly the most brilliant discovery at the peak of modern European civilisation, and an accomplishment that symbolised the even more prosperous future awaiting modern Europe. For the people of the nineteenth century, the ancient myths foretold a future of limitless development of human civilisation through governance over nature by the wisdom of man, and spoke of the strength of humankind and the wonder of love.

At the end of the twentieth century, however, the natural world has been exhausted by the prosperity of modern European civilisation. Humans are stuck on the point of coexisting with nature. Was it not at this time that the ancient myths and epics, written in an era when the great gods of the earth flourished, began to speak to modern people once again? Are they not ringing modern people's alarm bells, and hinting at the future path to be chosen? Umehara Takeshi's *Gilgamesh* (Umehara, 1988) certainly explicates what the ancient tale says to modern people very poignantly. Reading the truth in the ancient myths and epics is only possible by a minority of people of great sensitivity and intuition, like H. Schliemann.

The oldest human myth was written in Mesopotamia, 5,000 years ago, when humans first developed an urban civilisation (see Chapter 1). That epic was Gilgamesh. The forest god Humbaba is described in Gilgamesh. However, Humbaba's end was pitiful. The half-god, half-beast Humbaba, commanded by the Sumerian god Enlil, guarded the forests of Lebanese cedar, towering above the branches, for several thousands of years. The sacred forests were profaned by the maelstrom of human desire, and Humbaba strove to prevent their destruction. However, one day the king of Uruk, Gilgamesh, came with his mighty axe in hand. King Gilgamesh seemed to have forgotten himself momentarily when faced with the beauty of the Lebanese cedar forests, but immediately thereafter he began to cut down these sacred trees that were the dwellings of the gods. The enraged Humbaba screamed at him in a voice like thunder, and breathing fire from his mouth, attacked King Gilgamesh. However, Enkidu, who had accompanied King Gilgamesh, was a formidable opponent. Eventually Humbaba lost his head. The mighty god of the forests, Humbaba, was destroyed by the civilisation-reared human Gilgamesh in this 5,000 year old story written when humans first began urban civilisation. The birth of urban civilisation marches directly in step with the killing of the forest gods.

Over 3,000 years after the forest god Humbaba was slain by Gilgamesh, a tale of the god of the forest was recorded in the *Nihon Shoki*. The god's name was Itakeru-no-mikoto. The *Nihon Shoki* says, 'Before this, when Itakeru-no-mikoto descended from Heaven, he took down with him the seeds of trees in great quantity. However, he did not plant them in the land of Han (Korea), but brought them all back again, and finally sowed them throughout the Great Eight-island-country, beginning with Tsukushi producing green mountains. That is why Itakeru-no-mikoto was styled Isaoshi no Kami. He is the Great Deity who dwells in the Land of Kii.' When Itakeru descended from heaven, he brought with him many tree seeds. He did not sow any in Korea, however, but brought them all to Japan. Then, starting out from Tsukushi, he scattered the seeds throughout all of the eight great Japanese islands and greened the mountains. He came to be called Itakeru because of this great triumph, and was revered at Kii-no-kuni (modern-day Wakayama Prefecture) as a highly powerful god.

While Humbaba and Itakeru were both gods of the forest, their final end is very different. My heartstrings were tugged by the contrasting endings of these two forest gods. The differences in the endings for these two gods of the forest, set out in the two epics, effectively reflect the history of forest civilisation both at that time and after. In Mesopotamia, which cut down its god of the forest, Humbaba, every last forest completely disappeared and eventually their civilisation was also destroyed. The Japanese islands, however, which venerated Itakeru as a successful god who scattered tree and plant seeds, were covered in dense green forest and their civilisation prospered.

The two myths written in an era when the gods of Nature still overwhelmed humankind hint at the two completely different histories that humans were to walk thereafter. These are the histories of deforestation and forest civilisation. This work was planned to elucidate the unique characteristics of Japanese civilisation by tracing the histories of these unique relations between forests and civilisation, and by using comparative research, to determine the role of Japanese civilisation in this era of global environmental crisis.

■ WORLD DOMINATION BY DEFORESTATION CIVILISATION

The deforestation civilisation I speak of here does not just mean civilisation in the narrow sense that it developed in north-western Europe. Here, I call 'deforestation civilisation' the civilisation that was born in western Asia around 5,000 ^{14}C years BP, the civilisation that in a narrow sense, developed in the Mediterranean region in Europe north of the Alps from the twelfth century onwards. Based on geographical discoveries from the fifteenth century onwards, the civilisations diffused and scattered from the ancient continents of Asia and Africa to the new worlds of America and Australia, and eventually came to dominate the world at large.

This deforestation civilisation is one of hierarchical dominance, founded upon wheat cultivation and the conquest of nature. It is a civilisation founded upon human-centrism. Just as if an amoeba had enveloped the world, the civilisation of deforestation has dominated the earth within a period of five thousand years. In consequence, people in the late twentieth century are now directly confronted by a major crisis over peaceful coexistence with the global environment.

■ THE BIRTH OF A CIVILISATION OF HIERARCHICAL DOMINANCE

The starting point for a civilisation of hierarchical dominance and deforestation goes back 12,500 years, going by ^{14}C dating. During that era, the end of the long ice age was fast approaching, and the temperature was changing to that of the milder post-glacial age. With the climatic increase in temperature and humidity, the plains on which the great mammals such as the woolly mammoths, bison, and horses lived shrank, and the forests of deciduous oak and pines expanded. As their living environment worsened, great mammals declined in number. This was spurred on by human over-hunting. The people of the late Palaeolithic left behind the great plains from which their principal source of food, the great mammals, had disappeared, and began to live by strongly depending on the resources of the forests, which had spread anew. In addition, they started to live permanently both within the forest and on its fringes.

However, about 11,000 ^{14}C years BP, these one-time permanent dwellings were once again threatened by the drying up of forest resources due to the Younger Dryas cold period, and people were once again faced with a food crisis. With the receding of the forests and advancing desertification caused by the return of the cold of the Younger Dryas, the people of western Asia turned their attention to herbage of Gramineae. Barley and wheat grew wild on the great plains. By turning to the cultivation of wild grains, humans were able to survive the food shortage. Wheat farming, however, which was conducted alongside sheep and goat husbanding, produced a much greater surplus of foodstuffs than had been anticipated when the farming first began. Because of the harsh climate of the plains, it was indispensable to stock surplus to survive. In order to survive yourself, the only available option was to exploit other people. Fortunately, the staple grains of barley and wheat were ideal for long-term, storage unlike the meat of the large mammals.

In this way, the concepts of personal possessions and acquisition of wealth were born, and humans learned the joys of owning property and ruling over others. It was at this point that the path was opened leading to the development of a hierarchical civilisation in which a small handful of the wealthy and powerful ruled over everyone else. The amplifying effect of such limitless cruelties as governance by force, exploitation, and genocide had its genesis in the birth of wheat farming. Kawai (1979) calls this the 'Devil's Programme'.

▪ THE BIRTH OF CIVILISATION TO RULE OVER NATURE

The great rulers appeared along with the birth of urban civilisation 5,000 years ago. It was the birth of nations and of kings to lead them. The era of the birth of these urban civilisations also coincides with fluctuations in the earth's environment. Five thousand years ago saw the end of the hypsithermal, considered to have been a period of optimal climate and the advance of an increasingly cold one. Alongside this drop in temperature, the lowlands of Mesopotamia grew increasingly arid. With this drying out of the climate and increasing desertification, people gathered alongside large rivers. This proved to be the starting point for the birth of leaders of great power and of urban civilisation (Yasuda, 1991).

Gilgamesh, king of Uruk, typifies these great leaders. Gilgamesh made Uruk a great city, and launched a programme of deforestation of the Lebanese cedars in order to build himself a palace. With the help of his friend Enkidu, King Gilgamesh killed the god of the forests, Humbaba, and took the Lebanese cedars for himself. This is the world's oldest tale, written in the world's oldest script. Thus, the world's oldest tale is actually one of deforestation. The meaning of urban civilisation is the birth of the conquest of Nature. Ever since then, the unbounded greed of great leaders has acted to ceaselessly spur on humans to destroy the natural world.

▪ THE BIRTH OF MONOTHEISM

Ideas approving of the human dominance of nature, and carrying this concept of hierarchical rule not only into human ethics, but even into nature itself, was born around 3,200 years ago.

This was the birth of a single God in human form. The era in which this single God in human form was born was also a period of fluctuation in the global environment. As a result of our pollen analyses (Yasuda, 1991, 1993) conducted in western Asia, it is clear that the climate began to become much colder from around 3,200 years ago. As a result of this freezing, Israel and Egypt south of 35° North latitude grew more arid. The water level in the Nile fell, causing greater desertification (Hassan, 1986).

Moses wrote about it in the Old Testament (Exodus 1:5). Unable to stand by and watch a colleague harshly treated, Moses demands the liberation of the Hebrew people from the Pharaoh. When the Pharaoh refuses, he is cursed by God, and the Nile is transformed into blood and there are eruptions of frogs, lice, and flies, and hail falls from the skies, and the livestock fall sick and die everywhere around. Hail falling in Egypt indicates a fall in climatic temperature, the story of the Nile turning to blood indicates the fall in water level due to the climate becoming more arid and turning into a river of red clay.

This clearly shows that climatological disasters occurred during Moses' time. Moses' Exodus is said to have been a tale from Egypt during the reign of the Pharaoh Rameses II (1304 BC to 1237 BC) of the nineteenth dynasty. Amidst the confusion caused by the worsening of the climate around 3,200 years ago, Moses escaped from Egypt, and went on to receive the Ten Commandments on top of Mt Sinai. Thus, the deep shadow of climatic change casts itself even over the birth of anthropomorphic monotheism and the promulgation of the Ten Commandments according to the God Yahweh.

Under the auspices of monotheism, nature came to be the humble provider of human food. This cosmology was founded upon the nature-conquering human-centric notion of man beneath God, and nature beneath man. At the same time, their own God Yahweh came to be the one and only absolute, and a powerful sense of hierarchy developed which viewed other religions as heresy. In addition, the concept of hierarchical rule extended not only to human ethics, but also to the cosmology of nature. The Old Testament clearly states that animals are divided into clean and unclean animals.

In this way, monotheism classified the beasts of nature into clean and unclean, and gave humans a special place in the hierarchy of nature. This idea of hierarchical governance was passed from Judaism to Christianity, and came to be applied to all relationships; not only person to person, ethnic group to ethnic group, and nation to nation, but also to the relationship between people and Nature.

■ THE SPREAD OF HUMAN-CENTRIC CIVILISATION

It was around the twelfth century that this nature-conquering, hierarchical civilisation began to envelop north-western Europe. This period saw the end of the warm age known as the medieval warm epoch. The reclamation of the great forests of north-western Europe advanced rapidly due to the temperate climate, the technological revolutions of the plough, water wheel, and windmill, and the establishment of the 3-field system of crop rotation, allowing a field to lie fallow once every three years. Ito (1993) calls this period the 'twelfth century Renaissance'. Fig. 1 shows the transition of a famous European forest by Darby (1956). North-western Europe north of the Alps was covered in lush forests of European beeches and Japanese oaks (Fig. 1a). However, since the major reclamations from the twelfth century onwards, the forests have disappeared rapidly (Fig. 1b).

At the forefront of this deforestation stood the Christian missionaries. Sweeping aside the traditional animistic gods of the Celts and Germans who lived in the forests, they destroyed the sacred groves and tore down the sacred trees. This was in order to open up the darkness of the

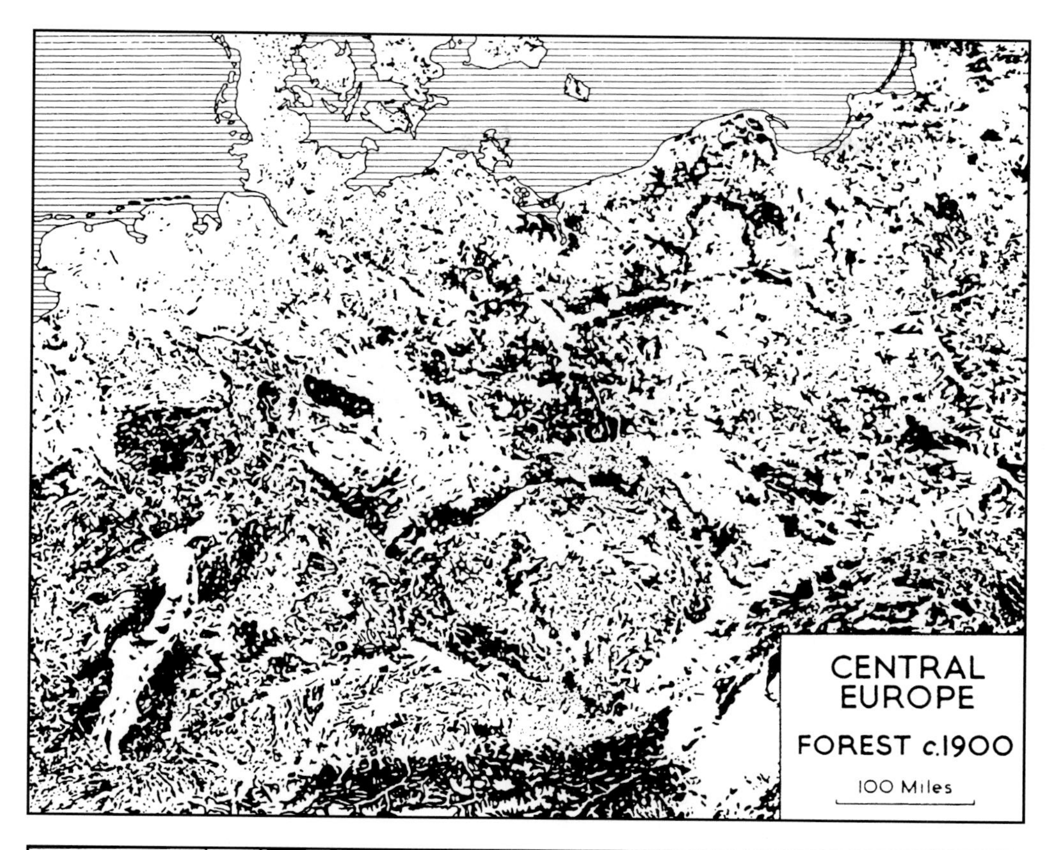

Fig. 1: The distribution of the forest in AD 900 in Europe (a) and in AD 1900 (b) (Darby, 1956)

Fig. 2: Bog man excavated from Tollund moor, Denmark (Silkeborg Museum)

forests for development, liberate heathens from their barbaric rituals of sacrifice, and introduce the shining light of human-centric civilisation.

The 'animal trials' that began in the twelfth century (Ikegami, 1900) were unique symbols of the permeation of Christianity into the forests where animism stubbornly lingered, and its increasingly expanding presence. The Druids, who governed the religion of the forests, sacrificed animals, and occasionally even humans, to maintain and restore the natural order. The bodies of over 600 bog people (Fig. 2), who were sacrificed to the great goddess of the harvests, have been discovered in the peat bogs of Denmark and northern Germany. Humans were offered up in sacrifice to maintain the order of Nature.

By contrast, the sacrifices for the 'animal trials' that began in the twelfth century were animals. Animals that had reduced order to chaos through bestiality and insects (see Chapter 11) that caused bad harvests, thereby destabilising human society, had to be put on trial and punished. These 'animal trials' sacrificed the animals to maintain the correct order of human society. The trials legitimised the sacrifices.

This indicates a reversal in the relationship between humans and Nature. It shows a shift from an era when humans were sacrificed to maintain and restore natural harmony, to an era in which animals and nature were sacrificed to maintain and restore the order of human society. This hierarchical civilisation, founded on nature-conquering human-centrism, spread throughout the world, following fifteenth century geographical discoveries.

By the fifteenth century, the forests of north-western Europe had already been violently destroyed, and farming and animal pasture had expanded greatly. In addition, this was an era in which climate cooling, called the little ice age, advanced rapidly. As the climate worsened, the plague raged throughout Europe and people faced a crisis. During this period, the number of witches dramatically increased (see Chapter 10). The witches were created by the bullying of people who faced the crisis. Large numbers of people abandoned the European continent, from which the forests had disappeared, and set out in search of the New World. First the African continent, and then the animistic civilisations of Asia and the New World fell one after another to the rapaciousness of human-centred deforestation civilisation. Again, the standard-bearer for this invasion was the Christian missionary. Precisely because people who believed in Christ had a hierarchical view that gave them a superior standing they went so far as to sell the Blacks of Africa across the water as slaves in the New World.

In this way, the world fell into the hands of the human-centrism, nature-conquering, hierarchical, deforesting civilisation whose spiritual foundations lay in monotheism. The civilisation of deforestation expanded and developed on the back of desertification to eventually envelop the world. Certainly, this civilisation brought about sterling materialistic development for humankind. At the same time, however, it spurred on a human history replete with 'eras of human exploitation and genocide,' and has resulted in an era of global domination that 'exploits nature for the advantage of humans alone.'

■ THE FIFTH DESERTIFICATION

With the first desertification, caused by the cold of the Younger Dryas around 11,000 ^{14}C years BP, the forest-destroying civilisations began wheat farming. With the second desertification caused by climate cooling at the end the period of optimal climate 5,000 years ago, deforesting civilisations gave rise to urban civilisations, and when the third desertification took place around 3,000 years ago, monotheism was born. Fifteenth-century Europe had already faced the exhaustion of forest resources. It was the people of Spain and Portugal who sought out the New World, and there were mass waves of migration directly motivated by the loss of forest resources, and because desertification was occurring in front of their eyes. The worsening of the climate in the little ice age exacerbated the problem. Let us call this the fourth desertification. It was during this fourth desertification that the global rule of forest-destroying civilisation began.

African and New World civilisations were invaded and destroyed in rapid succession on the pretext of their primitiveness and barbarism, yet it was these very 'primitive' and 'barbarous' civilisations that were, in truth, the civilisations of the forest. People worshipped the sacred trees and lived in a world of animism that looked upon the animals of the forest as the messengers of the gods. This was a world in which people lived in harmony with Nature and with the forest. This civilisation of the forest, which lived in peaceful coexistence with Nature, was scorned as primitive and barbarous, driven from its home, and buried in the darkness of history. Just how much the forests were destroyed by this invasion is revealed by the results of pollen analysis from the American continent. The native Americans were people of the forest who lived in harmony with their surroundings in the same way as Japan's Jomon people. Nevertheless, through the colonisation of the Europeans, who brought livestock with them, the forests of pine and deciduous oak were rapidly destroyed. America in the 1620s was a land of forests (see Fig. 3a). Three hundred years later, the great American forests have rapidly disappeared (see Fig. 3b) (Goudie, 1981).

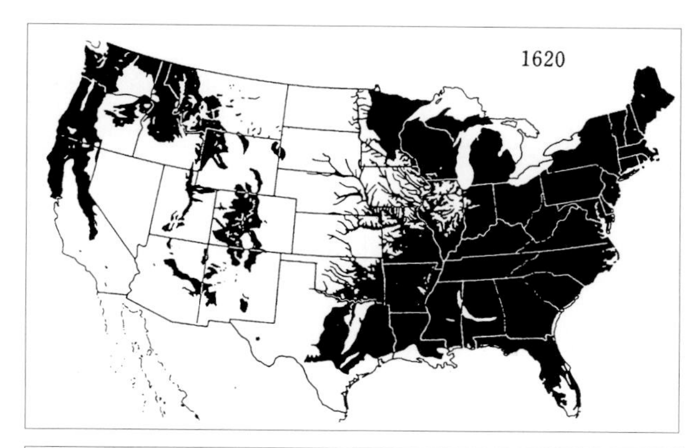

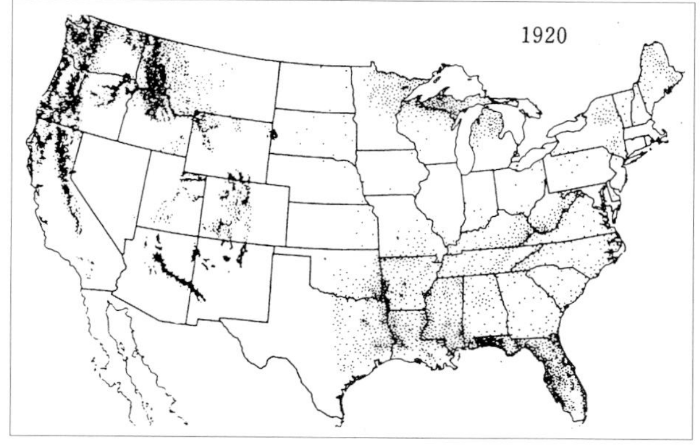

Fig. 3: The distribution of forest in AD 1620 in the USA (a) and in 1920 (b) (Goudie, 1981)

Fig. 4: A terrace field in Machu Picchu, Peru prevents soil erosion (Photo by Yasuda)

Fantastic stone buildings built by the Indio people are still to be found in the Maya city in Mexico (see Chapter 5) and Incan city of Cusco. Why, however, did they not build with wood? By contrast, vast quantities of lumber were used in the churches built following the Spanish invasion. The people of Indio largely constructed stone buildings and therefore the Inca civilisation was thought to be a civilisation of stone. That was before the forests had been swallowed up at the time of the Spanish conquest, and lush Warango trees covered the high plains of the Andes. I believe that, in all probability the stress of the Indio people on stone as the building material was because they knew that the protection of the forest ecology within the Andes mountains was a necessary and indispensable prerequisite to the maintenance of their civilisation. The terraced fields carved out of the ruins of Machu Picchu (Fig. 4), thought to reach up to heaven, expertly prevented soil erosion. In this lay the wisdom of the Indio people, who were wisely aware of the importance of natural ecology, and lived in harmony with Nature. The Spanish, however, destroyed this animistic civilisation that strove for peaceful coexistence with nature and also completely obliterated the forests that had been abundant during the Inca period.

In the latter half of the twentieth century, 500 years after the deforestation society spread and enveloped the world, humankind is confronting the fifth desertification. The cause of this desertification is not only deforestation; climate change, in the form of global warming caused by people, is also a major factor in the acceleration of desertification.

What eventually will be the impact of this fifth desertification on humans? This is extremely difficult to predict. Looking at the history of forests and civilisations to date, however, in eras when the forests have receded and desertification linked to climate change has been strikingly conspicuous, major revolutions have inevitably occurred that have engulfed all of humanity. It is difficult to believe that this desertification of the late twentieth to the twenty-first century can alone be an exception. In all probability, as the author Yasuda predicted in 1995, a period of massive change in human history equal to or even greater than the urban and spiritual revolutions, will take place within the next hundred years. With the four great desertifications to date, humanity has avoided disaster by creating new technologies, new cosmologies, and new social systems. Confronted by the current desertification crisis, is it possible that a technological revolution, ideas that change the direction of the system of civilisation, will appear that can bring about a revolution in human history? So long as this does not happen, modern civilisation may well crumble into dust.

■ THE EARTH IN MINIATURE

Do you know what entasis is? The vivid swellings of the pillars at Horyuji Temple and Toshodaiji Temple have stirred up the romantic feelings of many Japanese researchers, not least of the celebrated philosopher Tetsuro Watsuji and Katsuichiro Kamei (1943). Watsuji (1919 and 1947), who visited Horyuji, wrote in 1947 that, 'The striking entasis of the columns of this building make us think of the constructions of ancient Greece, which excites an interest in us.' It is said that this entasis swelling was the result of the influence of Greek and Hellenic culture, which came along the Silk Road. The swellings of the pillars in the great marble temples of the Mediterranean extended their influence as far as the pillars of temples in the Asuka Period. The effects of Greek civilisation, the starting point of glorious modern European civilisation, even reached as far as Nara culture and besides being a faint source of admiration in Greek and modern European civilisations, it has found a permanent niche in the memories of people down through the ages.

The ruins of Knossos, which represent Minoan civilisation, are to be found in Heraklion on the island of Crete (Fig. 5). The Cretan Minoan civilisation came into being in the Mediterranean world around 2000 BC. What supported their sudden economic development were the abundant resources of the forest. The results of Aghio Galini's pollen analysis on Crete (Bottema, 1980) testify that the island of Crete at that time was verdantly covered in forests of deciduous oak and evergreen as well as pines. Crete was an island of forests. At the time, Syria and Mesopotamia were already facing the depletion of forest resources. Crete, covered in its forests, used its abundant natural resources, and was thus able to magnify its power as a seafaring trading nation in the Mediterranean. The reason was that matchwood was needed as fuel for smelting bronze, which was a valuable commodity at the time, and in the manufacture of earthenware pots of daily use. These were exported overseas, and wood was also required to build the trading boats. The Ancient Mediterranean civilisation could not have survived without the forests that Crete possessed.

The ruins of Knossos (Fig. 6) at Heraklion are famous for having been unearthed and restored by the archaeologist A. Evans (1851–1941). The pillars of the Knossos ruins display the puffed markings of entasis. Evans restored the pillars using stones and concrete, but they were originally made of wood. In addition, the pillars were painted red (Fig. 7). The ruins of Knossos were once a temple made of wood, and it is in the very swelling of the red wooden pillars that the roots of entasis are to be found. The entasis swelling of the chalk marble pillars of Greece are traces of an ancient temple made of wood. Mediterranean civilisation was once one of wood and forests.

Fig. 5: The island of Crete, Greece (Photo by Yasuda)

Fig. 6: The Knossos, Heraklion, Crete, Greece (Photo by Yasuda)

Fig. 7: Restored pillar by A. Evans using stones and concrete, in Knossos, Crete, Greece (Photo by Yasuda)

There were however limits to the resources of these island forests. Even this former kingdom of forests, at the end of its civilisation's life, fell to the point where it was forced to import timber from Messina on the Peloponnesian Peninsula, and from the west coast of Turkey. The shortage of timber not only caused a shortage of building materials and fuel for the palaces, but also made it impossible to build the boats that maintained the economic arteries of the seafaring Minoan state. Besides, the deforestation also caused soil erosion causing a fall in grain production. At the end of the Minoan period there was a drop in the quality of their civilisation, and skeletons have been discovered that imply famine and cannibalism.

■ THE TRAGEDY OF EASTER ISLAND

In exactly the same way, but over than 3,600 years later, the same thing happened on Easter Island (Fig. 8), a small volcanic island in the Pacific offshore from Chile in South America. On this lone volcanic island there are over a thousand enormous stone Moai (Fig. 9), enormous statues that stand 20 m high and weigh up to 90 tons. The people who built these giant stone statues were Polynesians who came to the island from the west around AD fifth century (Fig. 10). They first started to build the enormous stone statues around AD tenth century. The stone Moai were built using stone tools to shave soft tuff (Fig. 11). These were then carried to the coast some 10 km away and erected there facing inland (Fig. 12), where they could watch over the village. These giant stone Moai were said to be ancestral gods.

However, somewhere around the mid-seventeenth century the construction of these enormous stone Moai was suddenly abandoned (Fig. 13). Over 200 unfinished Moai were simply left in the Rano Raraku quarry (Fig. 14), and others were simply abandoned en route to their

Fig. 8: Easter Island. We cannot see any forest apart from the Eucalyptus planted in the twentieth century (Photo by Yasuda)

Fig. 9: Stone Moai at Akivi (Photo by Yasuda)

Fig. 10: Present-day islanders (Photo by Yasuda)

Fig. 11: The author stands in front of the Moai carved in soft tuff in Rano Raraku (Photo by Yasuda)

Fig. 12: Moais are standing at Tahai keeping a watch inland (Photo by Yasuda)

Fig. 13: The toppled Moais facedown at Akabanga (Photo by Yasuda)

Fig. 14: Giant Moais on the outer slope of Rano Raraku quarry (Photo by Yasuda)

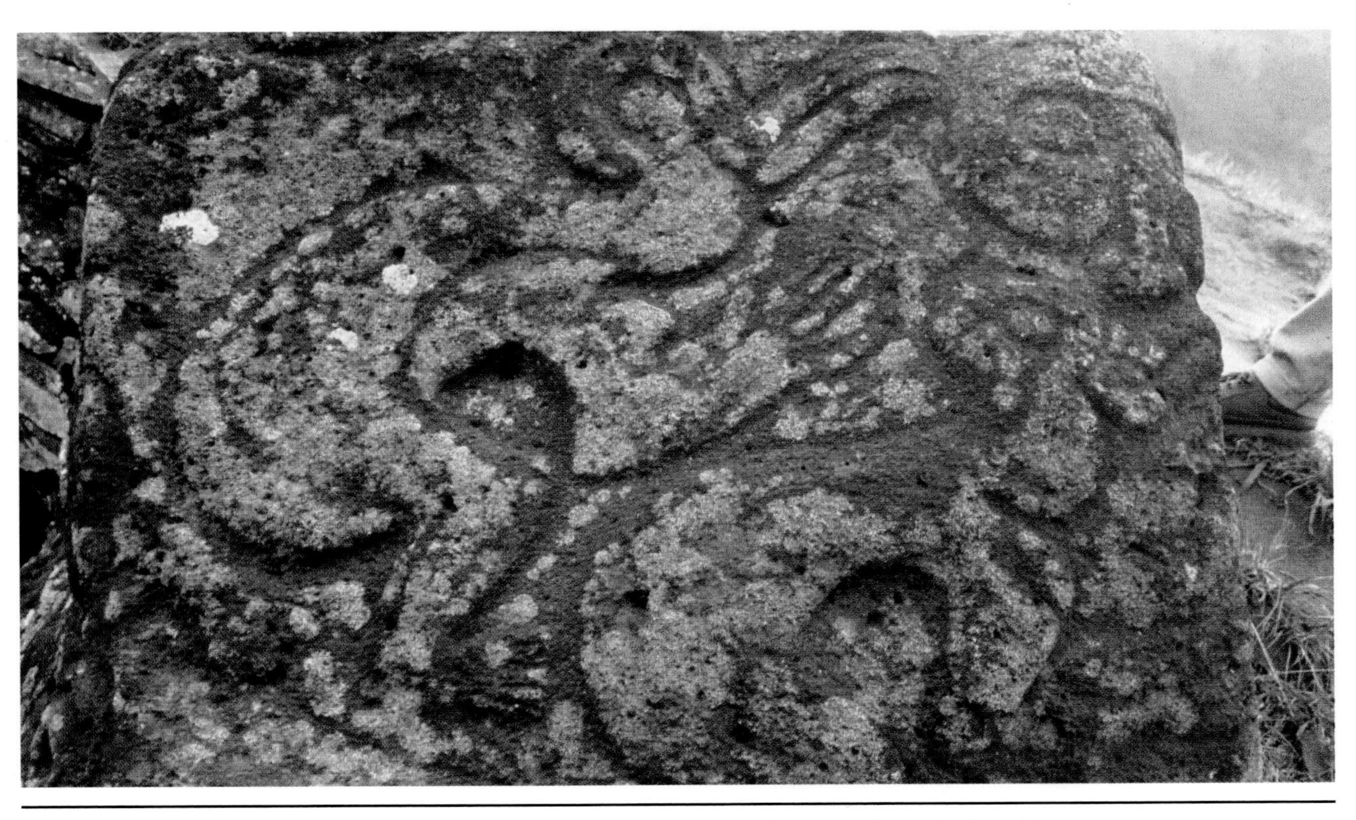

Fig. 15: The birdmen carved on rock at Orongo (Photo by Yasuda)

erection site. Why did this happen? Until now, the cause has been viewed as being an invasion by a new people who worshipped birds (Fig. 15), and famine and war.

Recently, however, a new theory has been presented (see Chapters 3 and 4), which traces the cause to the drying up of woodland resources. Professor John R. Flenley of Massey University in New Zealand, analysed the pollen in the sediment in the crater lake (Fig. 16) on the island (Bahn and Flenley, 1992). The results clearly show that from around AD seventh century immediately after people started inhabiting Easter Island, the coconut pollens, which had evidenced high levels until then, began to decline. In addition, in the seventeenth century, the island's forests had almost completely disappeared. As if to reflect this, the chicken coops and human dwellings made during this period were all of stone (Fig. 17).

Fig. 16: Crater lake Rano Kau (Photo by Yasuda)

Fig. 17: Very small entrance of a stone house in the ceremonial village of Orongo (Photo by Yasuda)

Wood was necessary to transport and erect the Moai. With the disappearance of the forests, the people were unable to move or erect their megaliths any longer. Besides, the deforestation accelerated soil erosion (Fig. 18), causing soil deterioration bringing about a reduction in the production of the inhabitants' staple crops of bananas and taro potatoes. Moreover, because trees were in short supply, the Easter Islanders became unable to even build boats for fishing. This in turn caused a food shortage, and the island was overcome by famine.

■ REPEATED MISTAKES

The forest resources of the island were clearly limited. The Minoan civilisation and the Moai civilisation of Easter Island were both destroyed when they had devoured every last scrap of their respective islands' forest resources. 3,600 years after the Minoan civilisation was destroyed, at the time when the Moai of Easter Island were discovered by the English explorer, Captain Cook, the forests of the British Isles from which he came had also already been virtually destroyed. In addition, scarcely a hundred years after the fall of the Moai on Easter Island, the same thing occurred on Minami Daito Island, Okinawa Prefecture, Japan. Reclamation and development since the Meiji Restoration has led, in virtually the blink of an eye, to the destruction and ultimate disappearance of forests of *Livistona* and banyan (Kuroda, 1996).

Fig. 18: Surface soil in the Easter Island was eroded by the heavy rain (Photo by Yasuda)

Today we should sense that this planet on which we live is one day very soon going to end up like Crete, Easter Island, or Minami Daito. The limits to the resources of planet Earth, floating all alone in the vastness of space, are clearly visible. When we humans have finished utterly devouring the finite resources of our planet, we will face the destruction and extinction experienced by the Minoan civilisation on Crete, the Moai in Easter Island, and of Minami Daito, which has become a sparsely wooded island. Crete, Easter Island, and Minami Daito are the future of this planet in miniature.

■ A FOREST CIVILISATION FLOURISHED ON THE JAPANESE ARCHIPELAGO

Japan is an archipelago of forests. It is extremely rare in the temperate zone to find an island nation so blessed with abundant forests. The climate, so well endowed with high precipitation, has brought about ideal conditions for nurturing forests. The average annual precipitation on the Japanese archipelago is extremely high at 1,700 mm. Moreover, rain falls evenly both in summer and winter, the precipitation even exceeding 4,000 mm in places such as Yaku Island and Odai-ga-hara.

Why then is it that the Japanese archipelago receives so much rain? The reason is because Japan is an island nation. The basis of Japan's heavy rainfall in summer is the evaporated water carried to the islands by tropical low pressure and tropical oceanic air masses that form over the Pacific. Moreover, the cause of the heavy snows that fall on the Sea of Japan side of the country in winter is evaporated water that forms over the Sea of Japan. That the country is surrounded by ocean is the greatest contributory factor in the creation of this forested topography. However, this forest archipelago has also changed greatly in the past 20,000 years, because of changes in the Earth's climate.

■ PALAEOLITHIC JAPANESE FOREST CULTURE

20,000 years ago, the Earth was at its coldest of the last-glacial age. The mean annual temperature of the Japanese archipelago was 7° to 8° C lower than it is today. That is why the sea level was also

Fig. 19: Reconstructed Palaeolithic landscape at Tomizawa site, Sendai City, Japan (after Morita, 1994)

at least 100 m lower, the Seto inland sea forming a land bridge, and the Japanese islands so different from what they are today as to be unrecognisable. Because of the low temperature, Hokkaido was partly covered in forest tundra. The region from Tohoku to Chubu was covered in sub-arctic coniferous forests. Western Japan was covered in a mosaic of deciduous broad-leaved trees, principally deciduous oak and conifers such as the Korean pine (*Pinus Haploxylon*) and Japanese hemlock. With the exception of the Kanto plains, which were spread with Graminae and mugwort due to falling volcanic ash, and Hokkaido, which was partly covered in tundra, the archipelago was covered with forests.

Most of the world's Palaeolithic cultures developed against a background of Graminae and Compositae. Their people were big game hunters who tracked horses and bison (Fig. 19). By contrast, Japanese Palaeolithic culture was one of forest hunter–gatherers (Yasuda, 1997). Besides hunting the indigenous deer (*Megaloceros*) that lived between the forests and the plains (Fig. 19), they also used such resources of the forest as the nuts of the Korean pine. Partially sharpened stone axes have been discovered all over the Japanese archipelago, and these are viewed as tools used to exploit the resources of the forest.

■ JOMON FOREST CULTURE IN JAPAN

The earth's climate warmed up and grew more humid rapidly about 12,500 ^{14}C years BP. As

Fig. 20: Reconstructed Jomon village in Togariishi site, Nagano Prefecture, Japan (Photo by Yasuda)

climatic humidity intensified, the forests of beeches, mainly centred on the Sea of Japan side, began to expand and grow. It is from these temperate deciduous forests that the oldest Jomon pottery was made. These comprised vessels for boiling the resources gathered from the forest, and it was through the discovery of vessels that the Jomon people were safely able to eat warm food from the forest and sea. Jomon culture adapted to the ecology of the temperate deciduous broad-leaved forests, giving rise to a forest-based culture that was strongly dependent on forest resources.

Between 6,500 and 5,500 ^{14}C years BP, the climate was at its warmest, most ideal, called hypsithermal, which was ideal. The mean annual temperature was 2° to 3° C warmer than today. Because of this rise in temperature, the sea level also rose, and the regions of Japan's present day major cities such as Tokyo, Osaka, Sapporo, and Niigata, which are all sited on coastal plains, were submerged. Due to this rise in temperature the lowlands of western Japan were covered with broad-leaved laurel forests of evergreen oak and chinquapin. In eastern Japan, the temperate deciduous broad-leaved forests of Japanese deciduous oak and chestnut expanded (Fig. 20). The early Jomon people used as food the nuts and berries of these deciduous oaks, chestnuts, and chinquapins, as well as the wild boar and deer from the forests and, additionally, freshwater fish such as salmon and trout, and the shellfish that lived in the inlets and bays. Research on bones from the Torihama shell

Fig. 21: Traditional Japanese landscape. The farmer's house is located between the *satoyama* and rice paddy field (Photo by Yasuda)

mounds in Fukui Prefecture has also revealed that the Jomon people ate tuna and bonito, and occasionally, whale (Morikawa and Hashimoto, 1994).

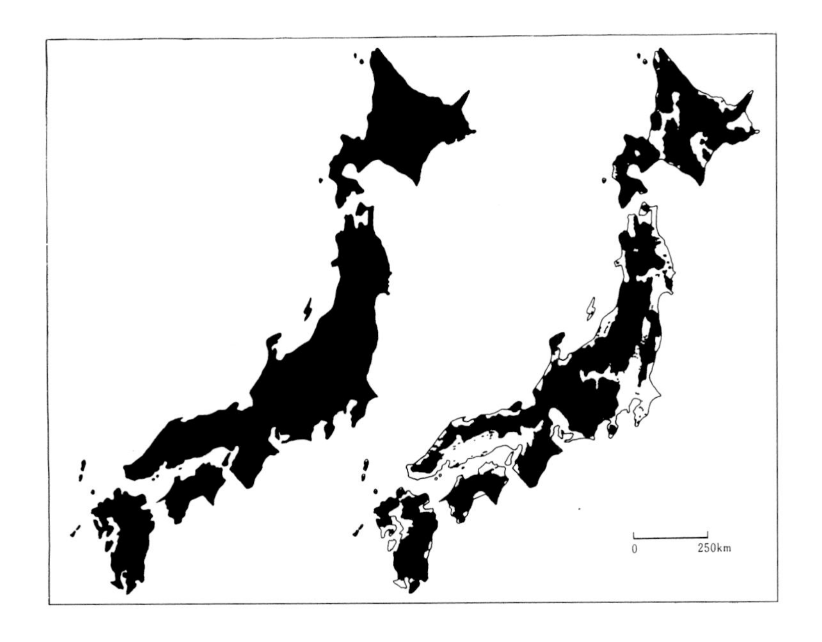

Fig. 22: Distribution of the forest at 3000 BP (left) and AD 1970 (right) (Yasuda, 1996)

■ *SATOYAMA* AND JAPANESE AGRICULTURE

Approximately 3,000 years ago, there was a sharp drop in temperature. As it fell, Japan's

Jomon culture faced a crisis. The deteriorating climate also saw an era of upheaval on the Chinese mainland. Different ethnic tribes amongst the local inhabitants moved from the north and west to the south and east. The upheaval of the Yin and Chou rebellions and the Spring and Autumn Annals Warring States Period were caused by the great influx of these ethnic refugees. Fleeing these social upheavals, people bringing rice and metal implements with them, arrived as boat people on the Japanese archipelago (Yasuda, 1992). However, these rice farmers, who initially brought livestock with them, stopped husbanding these animals, and switched to fish for their protein. Sheep and goat livestock ate to depletion the young buds of trees causing major deforestation. The Japanese succeeded in creating an agricultural society in the forest unburdened by the livestock that devastated the surrounding forest.

Japanese agricultural society was greatly dependent on the resources of the *satoyama* (see Chapter 12), until the spread of chemical fertiliser. It is these *satoyama* (Fig. 21) that symbolise forest-based agricultural society, and also testify that people and the forest lived symbiotically with each other. At the same time, the *satoyama* also provided the wild animals such as foxes and raccoon dogs with a habitat of their own. Agricultural society, striving for symbiosis with the forest, also realised a world of peaceful coexistence with the animals that inhabit it.

In this way, the people of the Japanese islands have been deeply involved with forests since Palaeolithic times. Ever since the Jomon Period in particular, the lifestyle of the Japanese islanders has been governed by forest time, to the extent that theirs can be called a forest civilisation. Ever since the Palaeolithic age, the types of forest have been changing. In the Japanese archipelago, however, the forest-based topography has continued unchanging. Fig. 22 shows the distribution of forests in the Japanese islands both ca. 3,000 years ago at the end of the Jomon Period, and today. Even though the type of forest has changed, the area covered by forest has clearly not changed as greatly as it has in Europe (Fig. 1) or America (Fig. 3). Japanese culture is the culture of the forest, and that culture is very long lasting. The secret of this perpetuity is, I believe, the continuation of this forest-based topography throughout the Japanese islands.

The keywords of forest civilisation are symbiosis, renewal, and rebirth, and egalitarianism in the relationship between humans and nature. I personally believe that, ever since the Jomon Period (indeed, it would not be an overstatement to say ever since Palaeolithic times), the key to avoiding the crisis of the fifth desertification and to rescuing the people of the future has lain in the traditions of forest-based culture, which has been built up assiduously by the people of the Japanese archipelago.

References

Bahn, P. and Flenley, J. (1992): *Easter Island, Earth Island.* Thames & Hudson, New York, 240 pp.

Bottema, S. (1980): Palynological investigations on Crete. *Review of Palaeobotany and Palynology*, 31: 193–217.

Darby, H. C. (1956): 'The Clearing of Woodland in Europe' in William L. Thomas, Jr. (ed.): *Man's Role in Changing the Face of the Earth*, Vol. 1., University of Chicago, Chicago, 1956, pp. 183–216.

Goudie, A. (1981): *The Human Impact on the Natural Environment*, Basil Blackwell, Oxford, pp. 38–39.

Hassan, J. (1986): Holocene Lakes and Prehistoric Settlements of the Western Faiyum, Egypt, *Journal Archaeological Science*, 13: 483–501.

Ikegami, S. (1990): *The Animal Trials: Justice and Cosmos in Medieval Europe. (Dobutsu Saiban)*, Kodansha, Tokyo, 234 pp.

Ito, S. (1993): *Renaissance in the twelfth Century: Influence of Arabian Civilisation on Europen World. (Juniseiki Runesansu: Seiou Sekai e no Arabia Bunmei no Eikyo)*, Iwanami, Tokyo, 227 pp.

Kamei, K. (1968): *Old Temple in Yamato: A Journey to Classical Beauty. (Yamato Koji Fubutsushi: Kotenbi e no Tabi)*, Obunsha, Tokyo, 1968, 350 pp.

Kawai, M. (1979): *Forests Gave Birth to Monkey: Contemporary*

Natural History. (Shinrin ga Saru o Unda: Genzai no Shizenshi) Heibon sha, Tokyo, 255 pp.

Morikawa, M. and S. Hashimoto (1994): *A Shell Mound at Torihama: A Time Capsule to Jomon Era. (Torihama Kaizuka; Jomon no Taimu Kapuseru).* Yomiuri Shinbunsha, Tokyo, 206 pp.

Morita, Y. (1994): Palaeoenvironment Reconstructed by the Fossil Pollen (Kafunkaseki ga kataru kokannkyo), Shukann Asahi Hyakka, Dobutsutachi no Chikyu, 138: 182–3.

Umehara, T. (1988): *Gilgamesh (Girugameshu)*: Shinchosha, Tokyo, 273 pp.

Watsuji, T. (1947): *A Pilgrimage to Old Temples (Koji Junrei)*. Iwanami, Tokyo, 287 pp.

Yasuda, Y. (1991): Climatic Changes at 5,000 Years BP and the Birth of Ancient Civilisations, *Bulletin of Middle Eastern Culture Center in Japan*, IV: 203–218.

—— (1993): 'Natural Environment in Japanese Islands' (Retto no Shizen Kankyo), *Japanese Islands and Human Society. (Nihon Retto to Jinrui Shakai).* Iwanami Koza Nihon Tsuushi. Vol. 1, Iwanami, Tokyo, pp. 43–81.

—— (1995): 'History Breakdowns: A Scenario of the Collapse of Modern Civilisation.' (Rekishi wa Keikokusuru: Gendai Bunmei Hokai no Shinario), in *Civilisation and Environment, (Bunmei to Kankyo).* Ito Shuntaro and Yasuda Yoshinori, (eds.), Nihon Gakujutsu Shinkokai, Tokyo, pp. 19–50.

—— (1996.): Forest and Civilisation *(Mori to Bunmei)*, Asakurashoten, Tokyo, 259pp.

—— (1997): *Environment of Jomon Civilisation. (Jomonbunmei no Kankyo).* Yoshikawa Kobunkan, Tokyo, 270pp.

Jomon dogu, 33.3 cm in height emphasising the breast, lower abdomen, genitalia, and eyes
(Utetsu Site, Aomori Prefecture. Photo by Tsuboi et al., 1977)

Forests were consumed by sheep and goats, Sultansazul Göl in Turkey (Photo by Yasuda)

Spring landscape of the alluvial plain in Min River, Schichuwan Province, China (Photo by Yasuda)

Contributors

Paul Bahn
428 Anlaby Road, Hull HU3 6QP, England

M. Barbero
Institut Méditerranéen d'Ecologie et de Paléoécologie, UMR CNRS 6116,
Faculté des Sciences et Techniques St-Jérôme, 13397 Marseille cedex 20, France

J.-L. de Beaulieu
Institut Méditerranéen d'Ecologie et de Paléoécologie, UMR CNRS 6116,
Faculté des Sciences et Techniques St-Jérôme, 13397 Marseille cedex 20, France

J. R. Flenley
Department of Geography, School of Global Studies, Massey University,
Palmerston North, New Zealand

Rosemary A. Joyce
Department of Anthropology, University of California, Berkeley,
California 94720-3710, USA

Jun Maki
Dept. of Parasitology, Kitasato University, School of Medicine, Sagamihara,
Kanagawa 228-0829, Japan

D. Marguerie
Laboratoire d'Anthropologie, UMR CNRS 153, Université de Rennes I,
35042 Rennes cedex, France

M. Reille
Institut Méditerranéen d'Ecologie et de Paléoécologie, UMR CNRS 6116,
Faculté des Sciences et Techniques St-Jérôme, 13397 Marseille cedex 20, France

H. Richard
Laboratoire de Chrono-écologie, UMR CNRS 9946, Université de Besançon,
25030 Besançon, France

André Siganos
Université Stendhal–Grenoble III–UFR des Lettres–BP 25 38040, Grenoble cedex, France

Hideo Tabata
Center for Ecological Research, Kyoto University, Kyoto 606-8502, Japan